Resistance to
Viral Diseases of Vegetables

Resistance to Viral Diseases of Vegetables

GENETICS & BREEDING

Molly M. Kyle
EDITOR

Timber Press
Portland, Oregon

Library of Congress Cataloging-in-Publication Data

Resistance to viral diseases of vegetables: genetics and breeding /
Molly M. Kyle, editor.
p. cm.
Includes bibliographical references and index.
ISBN 0-88192-256-0
1. Vegetables—Disease and pest resistance—Genetic aspects.
2. Virus diseases of plants. 3. Vegetables—Breeding. I. Kyle,
Molly M.
SB608.V4R47 1993
635′.0498—dc20 992-8777
CIP

Printed in Singapore

TIMBER PRESS, INC.
9999 S.W. Wilshire, Suite 124
Portland, OR 97225

Contents

Introduction

MOLLY M. KYLE

This volume is intended to cover the topic of genetics and breeding of resistance to viruses of vegetables, including aspects of the biology of the infection process and its interruption. It provides a critical synthesis of information from breeding programs and genetic studies that integrates developments in molecular biology and biotechnology. Plant viruses cause devastating losses to many major crops by reducing yield and quality, thus development and deployment of genetic resistance to this class of pathogens has been and will continue to be an important focus of a number of academic and commercial research programs. The relatively high value and strict market requirements for quality and appearance of vegetables contribute to the importance of developing resistance in commercially acceptable types. In contrast to other major groups of plant pests and pathogens, chemical control measures have not been effective against plant viruses. Genetic resistance, whether derived from naturally existing genes or engineered sequences, remains the only direct strategy to limit losses from these diseases.

An effort has been made to illustrate general principles and approaches while providing an appraisal of the status of research wherein viral resistance has been a significant breeding objective. This work has a long and productive history and is at the center of an explosion of interest because of the development and application of

Dr. Kyle is in the Department of Plant Breeding and Biometry, Cornell University, Ithaca, New York 14853.

new technologies. The outcome of these renewed and redirected efforts will depend on the degree to which they are integrated successfully with plant breeding programs. The inherent difficulty of this task, which involves bridging scientific disciplines and technical approaches, is compounded by the relative scarcity of published reports from either public or private viral resistance breeding efforts. The lack of common language between areas of research that increasingly focus on the same biological phenomena has hindered effective exchange of information.

The first chapters describe the process of identification of sources of resistance and their utilization in improvement of selected major vegetable crops. Cucurbit species are emphasized because considerable information is available from several long-term breeding efforts. These chapters also illustrate the increasing trend of the transfer of technology, theory, and breeding material from public to private workers for subsequent development, reflecting the changing relationships between public and private sectors. Anticipated impacts of the creation of novel sources of plant virus resistance and their use in vegetable improvement programs are discussed from both academic and commercial perspectives.

Subsequent chapters build upon this base by covering selected conceptual developments in breeding for viral resistance. The importance of multiple viral resistance has become clear with the release of varieties resistant to one disease that subsequently succumb to other diseases previously considered to be of secondary importance. The existence of simply inherited genes and gene clusters that confer resistance to members of the largest and most destructive group of plant viruses has been clearly established within the last five years. Two chapters discuss the genetic data leading to this conclusion and the implications for vegetable improvement, breeding strategies, and cellular and molecular studies of resistance mechanisms, resistance durability, and the evolution of the host-virus interactions.

Another significant conceptual development is the consideration of resistance that affects both disease severity, assessed at the organismal level, and disease incidence, which is a characteristic of the population. Resistance to vectors of viral disease and resistance that depends upon mechanisms that reduce transmission efficiency within a planting have received relatively little attention. Although high levels of field tolerance to viral diseases have been noted in some genotypes that clearly are susceptible to the virus, difficulty in selection and quantification of this attribute has slowed assessment of the potential benefits these plant characteristics could bestow.

Final chapters focus on the biology of the interaction between virus and host that results either in a susceptible or a resistant reaction. Plant viruses consist of a very few genes encoded by one or several nucleic acid molecules, usually RNA, enveloped by a protein coat. Despite a growing abundance of biochemical and molecular information about virus structure, very little is known about how these pathogens function to incite the drastic effects observed in susceptible hosts, or how resistance genes interfere with this process. These issues involve fundamental questions about host gene expression and viral replication, movement, and transmission. It is surprising that so little is understood about the mechanisms of pathogenesis and resistance to diseases that have been the object of direct scientific study for nearly 100 years.

Resistance and *susceptibility* are defined in each chapter in accordance with the authors' usage. This reflects fundamental disagreement about the meaning of even the most basic terms used to describe the host-virus interaction. Although this may appear to be a semantic issue, a number of specific examples show that imprecise communication among workers has hampered progress in this area of research. It is possible that when the molecular basis of viral disease expression and the capacity of a host to interrupt this process are understood, clear definitions will emerge; however, the difficulties stem as much or more from the difference in emphasis among pathologists, geneticists, and breeders, as from lack of understanding of the biological processes.

Among the issues that have proven troublesome for developing a terminology that is simple, consistent, and accurate is whether the trait in question is resistance to symptoms of the disease, i.e. the effects of the pathogen, or resistance to the pathogen *per se*. For a breeder interested in improving the performance of a genotype under field conditions, yield and quality of the marketable portion of the plant are of primary concern. For a virologist, the fate of the virus in a given genotype is of issue. Terms such as *immunity*, *hypersensi-*

tivity, and *tolerance* have been introduced to describe host reactions to plant viruses, but the specific relationships of these responses to the term *resistance* remain unsettled.

The biology of resistance genes also confounds attempts to draw clear distinctions. For example, alleles that determine certain and rapid plant death from the pathogen under some circumstances can occur at resistance gene loci, or can be due to conditional effects of the resistance allele itself. The Prv^2 allele in *Cucumis melo* and the *I* gene in *Phaseolus vulgaris* exemplify these types of responses (Ali 1950; Pitrat and Lecoq 1983). These reactions are not the occasional systemic breakdown of a "hypersensitive" resistant reaction, but literally extreme lethal sensitivity to some isolates of the pathogen. As mentioned above, resistance to the virus or a viral vector that is apparent at the population level with an effect on the epidemiology of disease has received considerably less attention than resistance obviously apparent at the organismal level. Finally, the development of standard methods to quantify resistant responses has not been a major area of research. Despite the existence of a number of alternatives, visual assessment of symptoms still remains the most widely used criterion of selection.

The genetics of resistance to plant viruses has been quite straightforward to determine in many cases. Many of the viral resistance traits of economic significance are conferred by a single gene. Contrary to widely held beliefs, the hypothesis that monogenically determined resistance is necessarily highly isolate-specific and therefore easily overcome by naturally occurring viral variants is not borne out by available evidence. The same allele, however, can show different inheritance to a given isolate depending on which phenotype conferred by the allele is assessed. Whereas the *bc-x* alleles in bean are truly recessive in their effects on symptoms, in some heterozygous combinations bean common mosaic virus replication is reduced significantly (Day 1984). Thus, the same genes are fully recessive when one phenotype, suppression of disease symptoms, is evaluated, and show gene dosage effects for a second phenotype, suppression of viral replication. Similarly, the *Tm-1* allele for resistance to tobacco mosaic virus in tomato is fully dominant and insensitive to temperature for suppression of mosaic symptoms, but when levels of viral replication are analyzed, the effect of the *Tm-1* allele is dependent on gene dosage and temperature (Fraser and Loughlin 1980, 1982). Molecular genetics offers much promise for the characterization of the genetic basis of viral pathogenicity and host resistance, and possibly for increased reliability in selection as well as for the creation and utilization of novel sources of genetic resistance.

One factor that will affect the success of more recently developed strategies is the extent to which engineered resistance genes are subject to the effects of "genetic background," as are naturally occurring genes. Background effects are especially prominent in wide crosses but can be quite dramatic between different genotypes within the same species. Evidence thus far suggests that regardless of the origin of expressed sequences, effects exerted by the rest of the genome and its interaction with environment are critically important. Both the inheritance of an allele and its phenotypic effect may be altered in the recipient genotype. The age of the plant at inoculation is also a highly significant variable in the expression of disease and can be manipulated in disease screening to provide the most favorable level of severity to discriminate among genotypes in segregating populations or advanced material. The effects of viral disease may become either more or less drastic if plants become infected closer to or at maturity. Environmental conditions, including but not limited to temperature, light intensity, light quality, and duration, also are well known to affect expression of resistance genes.

Although there is much activity in a number of areas related to host resistance, the relatively small size of plant viral genomes has allowed more rapid progress in the analysis of molecular processes involved on the pathogen side of viral pathogenesis. Molecular details of viral gene expression strategies, aspects of the infection cycle including replication, cell-cell and long distance movement, vector transmission, and regions of the viral genome involved in symptom expression are being elucidated for many plant viruses. Nevertheless, major gaps exist in our understanding of how plant viruses carry out their life cycles in the host, how these processes result in disease, and the ways in which the host interferes with viral pathogenesis.

Presently, a significant obstacle in the development of improved crops is lack of understanding of genetic variability among viral isolates. Confusion exists both about isolates or strains assigned to a single virus and about criteria used to distinguish closely related viruses that are members of the same virus family. In this volume, the term *pathotype* is used to denote groupings of viral isolates that may incite similar or different symptoms, but that are equally controlled by a resistance factor present in a host. It should be emphasized that this is a host-specific characteristic of viral isolates and may be valid only for one plant species or several species of the same botanical family. Consequently, the determination of a pathotype is a function of the genetic interaction between host and pathogen, and not solely a reflection of genetic variability of the pathogen. In this regard, viral pathotypes are comparable to physiological races of plant pathogenic fungi. It is likely that much progress could be made in a number of crops if clear host differential groupings and pathotype criteria were established, and collections of known pure isolates were characterized and maintained. These systems are also important when new isolates appear, as they offer a source for systematic searches of useful host responses that may be rapidly identified and deployed.

This volume has been made possible through the dedicated work of many people. The distinguished career of my teacher and colleague Dr. Henry M. Munger provided the inspiration to produce this volume under the auspices of the Vegetable Breeding Institute at Cornell University headed by Dr. Michael H. Dickson. As a glance at the list of authors and references will indicate, the success of many efforts undertaken around the world in a host of vegetable crops has depended upon the contributions of Dr. Rosario Provvidenti, Liberty Hyde Bailey Professor of Plant Pathology at the New York State Agricultural Experiment Station, Cornell University, at Geneva, New York. His careful observation, keen understanding of genetic variability of both plants and their viral pathogens, and his generosity have had a significant impact on progress in this area of research. The decision to undertake this work in its present form was made following a symposium organized with Dr. Donald H. Wallace in March 1989. I am grateful to the speakers and participants of that meeting, to the authors of the chapters included in this volume, and to my colleagues at Cornell University, especially Dr. Richard W. Robinson and Thomas A. Zitter, for critical reading of a number of chapters. The contributions of the reviewers and the copy editor at Timber Press, Suzanne Copenhagen, significantly improved the quality of this volume. The generous support I have received as the Burroughs Wellcome Fund Fellow of the Life Sciences Research Foundation during the years this book was produced is gratefully acknowledged. Finally, the many contributions of my husband Steven Kyle made it possible to accomplish this project.

Literature Cited

Ali, M. A. 1950. Genetics of resistance to the common bean mosaic virus in the bean (*Phaseolus vulgaris* L.). Phytopathology 40:69–79.

Day, K. L. 1984. Resistance to bean common mosaic virus in *Phaseolus vulgaris* L. Ph.D. Thesis. University of Birmingham, U.K.

Fraser, R. S. S., and S. A. R. Loughlin. 1980. Resistance to tobacco mosaic virus in tomato: effects of the *Tm-1* gene on virus multiplication. J. Gen. Virol. 48:87–96.

Fraser, R. S. S., and S. A. R. Loughlin. 1982. Effects of temperature on the *Tm-1* gene for resistance to tobacco mosaic virus in tomato. Physiol. Plant Pathol. 20:109–117.

Pitrat, M., and H. Lecoq. 1983. Two alleles for watermelon mosaic virus 1 resistance in melon. Cucurbit Genetics Coop. Ann. Rept. 6:52–53.

CHAPTER 1

Resistance to Viral Diseases of Cucurbits

ROSARIO PROVVIDENTI

Introduction

The Cucurbitaceae family comprises 118 genera with 825 species, which are widely distributed throughout tropical and subtropical regions of the world. Only relatively few of these species are extensively cultivated; the majority are wild and mostly confined to their native ecosystems. These species represent a great pool of genetic diversity that can be exploited for future use. A visit to the world's markets clearly demonstrates the importance of cultivated cucurbits in human nutrition. Some species of *Cucumis*, *Cucurbita*, *Citrullus*, *Lagenaria*, *Luffa*, *Benincasa*, *Sechium*, and a few others were domesticated thousands of years ago and, through the years, selected for better yield and quality. During the last century, cucumber, squash, melon, and watermelon have been subjects of active breeding programs. Similar to most all the other vegetables, cucurbits are affected by a large number of diseases caused by bacteria, fungi, viruses, and disorders attributed to genetic, physiological, or environmental factors (Whitaker and Davis 1962).

Diseases caused by viruses are often the most destructive and difficult to control. For years, chemical control of the vectors (mostly insecticides) and eradication of virus reservoirs (herbicides) have

Dr. Provvidenti is in the Department of Plant Pathology, Cornell University, New York State Agricultural Experiment Station, Geneva, New York 14456.

been the principal means of preventing viral diseases. As environmental issues have assumed greater importance in many countries of the world, efforts are increasingly directed toward minimizing the application of chemicals. These and other measures (mineral oils, reflective mulches or nettings, destruction of infected crops after harvesting, and strict sanitation for greenhouse crops) are still used, but they are only partially effective and must be repeated annually. In the last 60 years, efforts have been directed toward locating sources of heritable resistance. Time has demonstrated that genetic factors responsible for viral resistance are stable and durable, although they may be pathotype-specific. Resistance to viral diseases of cucurbits, as well as other crops, can derive from natural sources or from recent advances in biotechnology.

Traditional methods of control include (*a*) selection of resistant plants from existing varieties, primitive varieties or landraces, closely related species, or other genera from the same botanical family (resistance deriving from the first two sources can be quickly exploited and thus is more appealing; that originating from other species is often difficult to transfer because of genetic incompatibility or close linkage with undesirable characters); (*b*) genetics and breeding during which resistance genes (dominant or recessive, monogenic or multigenic) are characterized and then, through pedigree breeding or backcrossing, are incorporated into existing or new varieties; and (*c*) cross protection, which consists of inoculating plants with a mild or attenuated strain, thereby protecting them from destructive strains of the same virus.

Novel strategies for inducing viral resistance through genetic engineering include (*a*) complete viral genomes; (*b*) viral coat protein genes; (*c*) antisense sequences; (*d*) satellite sequences; (*e*) ribozymes; (*f*) putative viral replicase genes; (*g*) defective interfering RNAs; (*h*) antiviral antibodies; (*i*) expression of pathogenesis-related proteins; and (*j*) cloning of resistance genes. Most of the attempts to engineer viral resistance in cucurbits have involved the coat protein of the major viruses affecting these crops. This form of control is known as *coat protein–mediated protection*, or transgenic cross-protection, and is obtained by inserting the viral coat protein (CP) of a virus into plant genomes. Presently, other new ideas and approaches are being intensely pursued in several biotechnology laboratories. Each system offers advantages and disadvantages, while others remain to be tested (Grumet 1990).

Although a large number of viruses have been reported to infect cucurbits, those listed in Table 1 are considered of economic importance. These viruses have been grouped according to their vectors, which include aphids, beetles, fungi, leafhoppers, nematodes, thrips, whiteflies, and unknown biological agents. Although the majority of cucurbit viruses can infect a number of plant species belonging to different genera and families, a few seem to be restricted to members of the Cucurbitaceae.

Cucurbit Viruses Transmitted by Aphids

Aphid-transmitted viruses affecting cucurbits include bryonia mottle virus (BMV), cucumber mosaic virus (CMV), clover yellow vein virus (CYVV), muskmelon vein necrosis virus (MVNV), papaya ringspot virus pathotype (PRSV-W), telfairia mosaic virus (TeMV), watermelon mosaic virus (WMV), watermelon mosaic Morocco virus (WMMV), zucchini yellow fleck virus (ZYFV), and zucchini yellow mosaic virus (ZYMV). All these viruses are spread by one or more of a number of the following aphid species: *Aphis craccivora*, *A. glycine*, *A. gossypii*, *A. fabe*, *A. medicagins*, *A. spiraecola*, *A. rumicis*, *Acyrthosiphon pisum*, *A. kondoi*, *Aulacorthum solani*, *Hyalopterus atriplicis*, *Macrosiphum euphorbiae*, *Myzus persicae*, *Rhopalosiphum pseudobrassicae*, and others. All the viruses mentioned above are transmitted in the nonpersistent manner (style-borne) with varying degrees of efficiency. These viruses are usually acquired by aphid instars within one minute, but the aphids' ability to transmit them declines rapidly and is lost within a few hours. All of these viruses can be easily transmitted mechanically, and the use of phosphate buffers facilitates infectivity. Some (CMV, PRSV-W, WMV, and ZYMV) are widespread in temperate as well as warm regions of the world, where they often cause severe losses.

CUCUMBER MOSAIC VIRUS

Cucumber mosaic is a cucumovirus with icosahedral particles about 28 nm in diameter with a genome divided among three single-

Table 1. *Viruses affecting cucurbits.*

VECTOR	VIRUS	MODE OF TRANSMISSION
Aphids	Bryonia mottle virus (BMV)	Mechanical
	Cucumber mosaic (CMV)	Mechanical, seed
	Clover yellow vein (CYVV)	Mechanical
	Muskmelon vein necrosis (MVNV)	Mechanical
	Papaya ringspot-W (PRSV-W)	Mechanical
	Telfairia mosaic virus (TeMV)	Mechanical, seed
	Watermelon mosaic (WMV)	Mechanical
	Watermelon mosaic Morocco (WMMV)	Mechanical
	Zucchini yellow fleck (ZYFV)	Mechanical
	Zucchini yellow mosaic (ZYMV)	Mechanical, seed
Beetles	Melon rugose mosaic (MRMV)	Mechanical
	Squash mosaic (SqMV)	Mechanical, seed
	Wild cucumber mosaic (WCMV)	Mechanical
Fungi	Cucumber necrosis (CNV)	Mechanical
	Melon necrotic spot (MNSV)	Mechanical, seed
Leafhoppers	Beet curly top (BCTV)	
Nematodes	Tobacco ringspot (ToRSV)	Mechanical, seed
	Tomato ringspot (TmRSV)	Mechanical
Thrips	Tomato spotted wilt (TSWV)	Mechanical
Whiteflies	Beet pseudo-yellows (BPYV)	
	Cucumber vein yellowing (CVYV)	Mechanical
	Cucumber yellows (CYV)	
	Lettuce infectious yellows (LIYV)	Mechanical
	Melon leaf curl (MLCV)	Mechanical
	Squash leaf curl (SLCV)	Mechanical
	Watermelon curly mottle (WCMoV)	Mechanical
Unknown	Cucumber green mottle mosaic (CGMMV)	Mechanical, seed
	Cucumber leaf-spot (CLSV)	Mechanical, seed
	Cucumber pale fruit viroid (CPFV)	Mechanical
	Ournia melon virus (OMV)	Mechanical

stranded, positive-sense RNAs (RNAs 1, 2, and 3). The subgenomic RNA 4 that codes for the coat protein production is a duplication of the sequences present in RNA 3. CMV is often associated with small RNAs, called satellites, which usually contain 334–342 nucleotides that have little homology to the genomic RNAs of the virus. Satellites depend on CMV for their replication and are encapsidated in its particles. In some crops, satellites can greatly affect symptom development by the CMV genomic RNAs. CMV has been detected in the seeds of 19 other plant species, and apparently one CMV strain from India was able to be seed transmitted in muskmelon. This virus is of world-wide distribution and can infect about 800 plant species. In cucurbits, it can be very destructive by causing severe plant stunting, prominent leaf reduction, mosaic, and malformation. Fruits produced by infected plants usually remain small, distorted, and are often discolored and consequently unmarketable (Francki et al. 1979; Sharma et al. 1984).

Resistance to Cucumber Mosaic Virus

Cucumber Resistance to CMV in cucumber was found in 1928 in Oriental varieties 'Chinese Long' and 'Tokyo Long Green' (Porter 1928). From various inheritance studies conducted through the years, it can be concluded that resistance is conferred by three partially dominant factors (Kooistra 1969; Risser et al. 1977). Hence, in breeding for resistance, selection of plants with varying levels of resistance is possible, depending upon the number of genes involved. This resistance has had a great economic impact on the fresh market, as well as the pickling industry. Some biotechnology laboratories have also incorporated the coat protein of this virus (CMV-CP) in lines for commercial usage.

Melon Resistance to CMV is derived from Oriental melons, *Cucumis melo conomon*; and *makuwa*. Depending upon the source, resistance is governed by two or three complementary recessive genes (Karchi et al. 1975; Risser et al. 1977; Takada et al. 1979). Some of these factors for resistance appear to be strain and/or temperature dependent, hence plants develop symptoms at temperatures below 20°C (Risser et al. 1977; Stace-Smith 1984). Although resistance to this virus was reported in 1943 (Enzie 1943), only recently have private and state institutions been interested in developing resistant varieties. A few Chinese and Japanese varieties of pickling types are resistant or include a large number of resistant plants. The CMV-CP gene has also been introduced in some melon lines for commercial development.

Squash Most of the tolerance to CMV found in *Cucurbita pepo* appears to be polygenic and strain specific, hence of marginal value (Martin 1960; Lot et al. 1982). More effective resistance was located in wild species (*C. martinezii*, *C. ecuadorensis*, and *C. moschata* 'Nigerian Local') and transferred to summer squash varieties through interspecific crosses (Munger 1976; Provvidenti 1982; Washek and Munger 1983; Provvidenti 1990). This resistance appears to be partially dominant. Resistance has also been reported in some varieties of *C. maxima* from South America, South Africa, and Australia (Provvidenti 1982).

Watermelon Varieties of *Citrullus lanatus* are resistant to the most prevalent strains of CMV, except to a specific strain that can infect plants systemically, causing prominent foliar mosaic and fruit malformation. No attempts have been made to locate sources of resistance to the watermelon strain of CMV (Provvidenti 1986a).

Lagenaria Resistance was found in PI 269506 (Pakistan), PI 271353 (India), and PI 391602 (China) (Provvidenti 1981) and some accessions from Australia (Greber 1978). No information is available regarding the genetics of this resistance, but it can be strain-specific.

PAPAYA RINGSPOT VIRUS-W

Until a few years ago, papaya ringspot virus-W (PRSV-W) was known as watermelon mosaic 1 (WMV-1). A number of studies, however, determined that WMV-1 and PRSV are identical in many respects. Presently, two pathotypes of this potyvirus are recognized: PRSV-P (papaya strain), which infects *Carica papaya* and most of the Cucurbitaceae, and PRSV-W (watermelon strain), which infects all Cucurbitaceae but not papaya. For cucurbits, the latter pathotype is of great economic importance because of its destructiveness. PRSV-W is a member of the potyvirus group, hence its particles consist of flexuous rods about 780 × 12 nm containing a single strand of RNA. There are no reports of seed transmission. PRSV-W is common in the warm regions of the world, but occasionally it occurs in temperate zones. Symptoms incited by this virus include severe plant stunting, foliar mosaic, and extreme reduction of leaf lamina (shoe-string). Fruits are prominently affected by knobby overgrowths and color break (Purcifull and Hiebert 1979; Purcifull and Gonsalves 1984).

Resistance to Papaya Ringspot Virus

Cucumber Resistance and tolerance were found in a number of cucumber varieties from China, Japan, Hawaii, and South America (Provvidenti 1985). In the variety 'Surinam Local', resistance is monogenically recessive (*wmv-1-1* or *pr-1v*) (Wang et al. 1984; Pitrat 1990). Resistant varieties respond with transitory systemic symptoms confined to one or two leaves, but fruits are unaffected. A close linkage was found between *prv-1* and *bi*, a gene for bitterness that could be used as a marker (Wang et al. 1987). A new generation of cucumber lines carrying the resistance gene will be available shortly from several seed companies. Experimentally, PRSV-W coat protein (CP), has been introduced in some lines for commercial use.

Melon A good level of resistance in melon to PRSV-W was located in PI 180280 from India, in which it is conferred by a single dominant gene (*Wmv-1* or *Prv1*) (Webb 1979; Pitrat 1990). A second allele at the same locus, found in PI 180283, conditions a lethal hypersensitivity (*Wmv-1-2* or *Prv2*) (Pitrat and Lecoq 1983; Pitrat 1990), hence it is not of economic importance. A very high level of resistance to PRSV-W due to a single dominant factor (*Prv*) was reported to occur in an accession of *Cucumis metuliferus* ('Kiwano'), PI 292190, from South Africa (Provvidenti and Robinson 1977). It has not been possible to determine whether *Prv* and *Prv-1* are identical, because of genetic incompatibility between this species and *C. melo*. Both of these genes are very stable in the homozygous condition, but some of the commercial F_1 hybrids with *Prv1/+* tend to develop varying degrees of stem and foliar necrosis following mechanical inoculation. The PRSV-W coat protein gene (CP) has been introduced in some melon lines, for commercial development.

Squash A high level of resistance is available in *Cucurbita ecuadorensis*, *C. ficifolia*, *C. foetidissima*, and in *C. moschata* 'Nigerian Local' (Nigeria) (Provvidenti et al. 1978; Stace-Smith 1985). Attempts to transfer resistance into breeding lines of *C. pepo* are underway, but progress is slow because of genetic incompatibility among species. In the Nigerian squash, resistance appears to be recessive. Resistance to PRSV-W is very desirable in varieties of *C. pepo*, since PRSV-W is one of the major viral agents in the tropics. Attempts are also under way to introduce the PRSV-W coat protein gene (CP) into some breeding lines.

Watermelon No valuable sources of resistance or tolerance have been reported in *Citrullus lanatus* and other *Citrullus* species. Some landraces of these species seem to possess some field tolerance, which could be useful if it can be transferred through breeding (Provvidenti 1986b). The PRSV-W coat protein gene (CP) could be very useful, particularly in association with the field tolerance.

Lagenaria The following lines were determined to be resistant: PI 188809 (Philippines), PI 271353 (India), PI 280631 (South Africa), PI 288499 (India), PI 391602 (China), and the variety 'Hyotan' from Hawaii (Provvidenti 1986b). Resistance was also found in other varieties from India (Bhargava 1977). No studies have been conducted on the heritability of this resistance.

Luffa Resistance was reported in some varieties of *Luffa acutangula* and *L. aegyptica* (= *L. cylindrica*) from India (Bhargava 1977). Some other varieties of these two species from Taiwan appeared to be tolerant (Huang et al. 1987).

WATERMELON MOSAIC VIRUS

Watermelon mosaic virus was formerly known as watermelon mosaic virus 2 (WMV-2). It is characterized by flexuous particles about 760 nm long containing a single strand of RNA typical of potyviruses. No seed transmission has been reported for this virus. WMV is able to infect many species in the Cucurbitaceae and Leguminosae, hence it is present in the tropics, as well as in temperate and cool regions of the world. Symptoms incited by this virus are less prominent than those caused by other cucurbit viruses and involve green mosaic, leaf rugosity, green veinbanding, and chlorotic rings. Generally, affected leaves remain of quasinormal size, but fruit color and shape may be adversely affected (Purcifull et al. 1984).

Resistance to Watermelon Mosaic Virus

Cucumber Resistance was first reported in the variety 'Kyoto 3 Feet Long' and was subsequently found in a number of other Japanese and Chinese varieties. In the Japanese cucumber, resistance was demonstrated to be monogenically dominant (*Wmv*) (Cohen et al. 1971; Robinson et al. 1976).

Melon For many years, attempts to find valuable sources of resistance to WMV in *C. melo* were without success; however, Drs. H. M. Munger and M. M. Kyle of Cornell University are developing breeding lines that possess a high degree of tolerance, which derived from a combination of factors found in landraces from India (Karchi et al 1975; Pitrat and Lecoq 1984).

Squash A good level of resistance is available in *C. moschata* 'Nigerian Local', *C. maxima* 'Pai Yu' (China), and in the wild species *C. ecuadorensis*, *C. ficifolia*, *C. foetidissima*, and *C. pedatifolia*. In all cases, resistance appears to be inherited recessively (Provvidenti et al. 1978; Provvidenti 1982).

Watermelon Some accessions of the Nigerian 'Egusi' (*Citrullus colocynthis* PI 494528, PI 494532, and a few others) have been reported to be resistant to this virus (Webb 1977; Provvidenti 1986b). This wild species is fully compatible with the cultivated watermelon and is presently exploited in breeding for resistance. A number of landraces of *C. lanatus* from Africa seem to possess a certain level of field tolerance.

Lagenaria Resistance was found in PI 271353 (India), PI 391602 (China), G-24386 (California), and 'Hyotan' (Hawaii) (Provvidenti 1981). Resistance was also reported in other Indian varieties

(Bhargava 1977). No information is available on the genetics of this resistance.

Luffa *L. acutangula* PI 391603 from China and other lines of this species were reported to be highly resistant to WMV (Webb 1965; Provvidenti 1977). Resistance also was found in two accessions of *L. acutangula* from India (Provvidenti and Robinson 1977) and one of *L. aegyptica*, also from India (Bhargava 1977).

Benincasa Two accessions (PI 391544 and PI 391545) from China were found to be resistant (Provvidenti 1977). Further testing may reveal additional sources of resistance for WMV, as well as for other viruses affecting this species.

ZUCCHINI YELLOW MOSAIC VIRUS

Zucchini yellow mosaic virus was first reported almost simultaneously in 1981 from Italy and France, where it was named muskmelon yellow stunt virus (Lisa and Dellavalle 1981). In less than five years, it was known to occur in 20 countries on five continents, often causing devastating epidemics. Virus particles are flexuous, circa 750 nm long, and contain a single-stranded RNA. Although circumstantial evidence points to seed transmission of this virus, demonstrating this avenue of spread has been very difficult (Provvidenti and Robinson 1987). Two recent reports have indicated a low level of transmission in some varieties of *C. pepo* (Greber et al. 1988; Schrijnwerkers et al. 1991). Unquestionably, this is one of the most destructive viruses occurring in cultivated cucurbits. The virus incites very prominent foliage mosaic, severe malformation, and plant stunting. Fruit often develop knobby areas, resembling those caused by PRSV-W, thus, differentiating the symptoms caused by these two viruses is often very difficult. Serologically, ZYMV is related to WMV. A few strains and pathotypes of ZYMV have already been identified (Lecoq and Pitrat 1984; Lisa and Lecoq 1984; Provvidenti et al. 1984).

Resistance to Zucchini Yellow Mosaic Virus

Cucumber A number of Chinese varieties were determined to be resistant (Provvidenti 1985). Inheritance studies based upon a single plant selection of the variety 'Taichung Mou Gua' (TMG-1) have established that the resistance is inherited monogenically recessive (*zym*) (Provvidenti 1987). Resistant plants may develop a systemic veinal chlorosis confined to a few basal leaves with some strains, or are systemically resistant with others. In no case are fruits affected. Some laboratories have incorporated the ZYMV-CP gene into cucumber.

Melon Resistance in some plants of PI 414723 from India is conferred by a single incompletely dominant gene, *Zym* (Pitrat and Lecoq 1984). This resistance, however, is strain-specific, thus, other strains can infect plants with this gene (Lecoq and Pitrat 1984). Efforts to find other sources of resistance have been unsuccessful, but some single plants of landraces of *C. melo* from Afghanistan and India possess a certain tolerance. A mild strain of ZYMV has been used with success in cross protection trials conducted in France and Taiwan. The ZYMV-CP gene has also been incorporated in some experimental lines which are being developed for commercial use.

Squash No resistance has been found to this virus in *C. pepo* or close wild relatives. Conversely, resistance is readily available in some accessions of *C. moschata* from Nigeria and Portugal. This resistance is conferred by a single incompletely dominant gene, *Zym* (Munger and Provvidenti 1987; Paris et al. 1988). The only wild species offering resistance to ZYMV is *C. ecuadorensis*, which can be utilized for interspecific crosses with *C. maxima*. This resistance is also conferred by a single incompletely dominant gene, *Zym* (Robinson et al. 1988). A mild strain of ZYMV has been used with some success in cross protection trials conducted in France and Taiwan.

Watermelon A very high level of resistance was recently found in some landraces of *C. lanatus* from Zimbabwe (Provvidenti 1991). This resistance is confined to only one strain, ZYMV-FL (Florida) and is governed by a single recessive gene, *zym* (Provvidenti 1991). Since this strain is widespread in the continental US, this resistance is very useful, but it succumbs to other strains of the virus present in Europe, Asia, and the Middle East. A non-strain-specific resistance was found in some accessions of *C. colocynthis* from Nigeria (PI 494528, PI 494532, and a few others) and is presently used in several countries. This resistance was proved to be temperature-dependent, and it is best expressed at warmer temperatures (Provvidenti 1986b).

Lagenaria A line from India (PI 271353) is resistant to ZYMV (Provvidenti et al. 1984). Recent studies have demonstrated that this line is resistant to several strains from different parts of the world, hence, it appears not to be strain-specific (R. Provvidenti, unpublished).

Luffa Some accessions of *L. acutangula* respond with systemic infection but fail to develop symptoms, consequently it is likely that this species is highly tolerant to this virus (Hseu et al. 1985).

Other Cucurbit Potyviruses Transmitted by Aphids

BRYONIA MOTTLE VIRUS

The host range of bryonia mottle virus is largely confined to cucurbits. Serologically it is unrelated to PRSV-W, WMV, and a few other potyviruses. It was found in Morocco, where it primarily affected *Bryonia dioica*. Symptoms consist of prominent foliar chlorotic and necrotic spotting, plant stunting, and flower abortion. Apparently BMV is not seed-transmitted (Lockhart and Fisher 1979). In a study involving many Cucurbitaceae, only *Momordica charantia* was not infected by this virus (Lockhart and Fisher 1979).

CLOVER YELLOW VEIN VIRUS

Clover yellow vein virus commonly occurs in legumes, in which it causes severe symptoms. In nature, it has been found to infect summer squash, causing numerous chlorotic leaf spots in yellow-fruited varieties. Fruits are not affected, but seed production may be reduced (Provvidenti and Uyemoto 1973; Lisa and Dellavalle 1981). A number of cultivated and wild *Cucurbita* species were found to be resistant (Provvidenti et al. 1978).

MUSKMELON VEIN NECROSIS VIRUS

Muskmelon vein necrosis virus appears to be restricted to melons (*Cucumis melo reticulatus*, *inodorus*, and *chito*), in which it induces a distinct veinal necrosis in all but the apical leaves. Petioles also become necrotic, together with a necrotic cork-like streaking of the stems. In nature, this virus is able to infect a number of leguminous species, and it is related to red clover vein mosaic virus (Freitag and Milne 1970). No resistance has been reported.

TELFAIRIA MOSAIC VIRUS

Telfairia mosaic virus was reported from Nigeria and mainly infects the perennial cucurbit *Telfairia occidentalis*, known as the fluted pumpkin. This plant is used for its edible seeds and nutritious aromatic leaves. TeMV is serologically closely related to ZYMV and more distantly related to WMV and BYMV. In nature, this virus appears to be restricted to members of the Cucurbitaceae, causing foliar mosaic and severe distortion, chlorosis, fruit malformation, and severe plant stunting. In fluted pumpkin, seed transmission of TeMV ranges from 6 to 20% (Nwauzo and Brown 1975; Thottappilly and Lastra 1987; Anno-Nyako 1988). Through repeated selection, a number of immune and tolerant *Telfairia* lines were developed in Nigeria, which have increased foliage and fruit yields (Atiri and Varma 1983). At least one accession of *Cucurbita maxima* was reported to be systemically resistant (Nwauzo and Brown 1975).

WATERMELON MOSAIC MOROCCO VIRUS

Watermelon mosaic Morocco virus is not serologically related to PRSV-W or WMV, but it incites symptoms closely resembling those caused PRSV-W. These include plant stunting, foliar mosaic, green blisters, and extreme reduction of leaf laminae (shoe strings). Fruits show knobby areas, distortion, and color break (Fisher and Lockhart 1974). Initially, WMMV was found to affect cucurbits in North Africa (Morocco), but later was determined to be of common occurrence in all the main cucurbit producing areas of South Africa (Meer and Garnett 1987). Some sources of resistance to PRSV-W are apparently equally effective against WMMV (H. Lecoq, personal communication). *Luffa aegyptica*, *Cucumis metuliferus*, *Coccinia sessifolia*, and *Citrullus ecirrhosus* appear to be resistant to WMMV (Meer and Garnett 1987).

ZUCCHINI YELLOW FLECK VIRUS

Zucchini yellow fleck virus has been reported from the Mediterranean region. It infects squash and cucumber, causing pinpoint yellow flecks on the leaves and fruit malformation (Vovlas et al. 1981). No resistance has been reported in any of the affected species.

Cucurbit Viruses Transmitted by Beetles

Three cucurbit viruses are transmitted by beetles: melon rugose mosaic (MRMV), squash mosaic virus (SqMV), and wild cucumber mosaic virus (WCMV). In nature they are spread by the striped and spotted cucumber beetles (*Acalymna* spp. and *Diabrotica* spp.), which acquire the virus within 5 minutes and retain it for about 20 days. There is no evidence of virus multiplication in the vectors, but it can be detected in the regurgitation fluid, feces, and hemolymph. These viruses are mechanically transmissible, but only SqMV is seedborne.

SQUASH MOSAIC VIRUS

Squash mosaic virus is one of the major cucurbit viruses to be transmitted via seeds. Hence it can be found wherever cucurbits are cultivated. A comovirus with isometric particles about 30 nm in diameter, it contains a single strand of RNA. The natural host range of SqMV is limited to Cucurbitaceae, but experimentally it can infect 15 plant species of 11 genera. Among the cultivated cucurbits, melon and squash species are the hosts for which most of the infection can be traced to infected seeds. Two strains or pathotypes have been characterized: the melon strain (or Strain I), which causes prominent symptoms in melon but mild symptoms in squash, and the squash strain (or Strain II), which incites prominent symptoms in squash and mild ones in melon. These two strains are also serotypes; thus, they can be easily distinguished in immunodiffusion or ELISA tests. Both strains of the virus incite a variety of symptoms, including mosaic, ringspots, green veinbanding, and protrusion of veins at the foliar margin. Under certain environmental conditions, infected plants of *C. moschata* and *C. pepo* may develop prominent foliar enation (Campbell 1971; Nelson and Knuhtsen 1973).

Resistance to Squash Mosaic Virus

Cucumber Although SqMV can infect cucumber, foliar symptoms are usually mild and fruits are rarely affected. Because this virus has not caused noticeable losses in this crop, no research has been conducted to find sources of resistance (Provvidenti 1986).

Melon SqMV can be a serious problem in any kind of melon. The use of certified seeds has been and still remains the most effective method of minimizing outbreaks and the spread of this virus. No high level of resistance has been found in *C. melo* to the melon and the squash strains of SqMV. Some landraces from India, Afghanistan, and Pakistan appear to be tolerant and respond to the melon strain with mild systemic symptoms, but whether this tolerance will prevent seed transmission of the virus remains to be determined (R. Provvidenti, unpublished). Tolerance to the squash strain of the virus was located in PI 157080 from China and a few other landraces from the Indian subcontinent (Webb and Bohn 1962). Some laboratories are isolating the coat protein gene (SqMV-CP) to be used for transgenic cross protection.

Squash The squash and the melon strains of SqMV are seedborne in *Cucurbita* species, but all attempts to find sources of resistance in cultivated species have been unsuccessful. Conversely, tolerance was located in some wild species, such as *C. ecuadorensis*, *C. martinezii*, and *C. okeechobeensis*, which react to infection with mild and transitory systemic symptoms (Provvidenti et al. 1978). No information is available regarding the mode of inheritance.

Watermelon Strains of SqMV can infect watermelon and its wild relatives, but generally it does not cause damage of economic importance. No research has been undertaken to find sources of resistance (Provvidenti 1986b).

Lagenaria All the accessions of *L. siceraria* were found to be resistant to both strains of SqMV (Provvidenti 1981). Plants respond with small necrotic local lesions without systemic infection. Hence this species can be used as an assay host for SqMV and for eliminating this virus in a mixture with other cucurbit viruses.

Luffa An accession of *L. acutangula* (PI 391603) from China was found to be highly resistant to SqMV (Provvidenti 1977).

MELON RUGOSE MOSAIC VIRUS

A tymovirus, MRMV has isometric particles about 32 nm in diameter containing single-stranded RNA. It was isolated from melon and watermelon grown in Yemen. Serologically, it is related to other viruses of the same group (eggplant mosaic, belladonna mottle, Andean potato latent, and others), but is distinct from wild cucumber mosaic virus, turnip yellow mosaic, Desmodium yellow mottle, and a few others. Systemic symptoms consist of foliar angular chlorotic patches followed by intense rugosity. Leaves remain small and misshapen. Fruits often are severely malformed. Possible beetle vectors include *Epilachna chrysomelina*, *Asbecesta transversa*, and *Raphidopalpa foviecollis* (Jones et al. 1986). No reports are available regarding resistance to this virus in watermelon, melon, and other cucurbits.

WILD CUCUMBER MOSAIC VIRUS

WCMV is also a tymovirus possessing isometric particles about 30 nm in diameter. Serologically, it is related to turnip yellow mosaic and a few other viruses of the same group. Originally, it was isolated from wild cucumber (*Echinocystis fabacea*) affected by a chlorotic mottle. This virus can infect several cucurbits, including melon, summer squash, and watermelon, in which it causes a diffuse yellow mottle that becomes less conspicuous as leaves age. Other strains cause green blisters and a bright yellow mottle. Apparently, this virus is seedborne (Regenmortel 1972). *Cucumis sativus* is resistant to some isolates of WCMV, and some accessions of *Cucurbita moschata* appear to be resistant (Regenmortel 1972).

Cucurbit Viruses Transmitted by Fungi

Two cucurbit viruses are transmitted by fungi: cucumber necrosis (CNV) and melon necrotic spot (MNSV). These soil-borne viruses are transmitted by the zoospores of the chytrid fungus *Olpidium radicale* (*O. cucurbitacearum*), which carries the virus particles externally. They are also mechanically transmissible, and MNSV is seed-borne.

CUCUMBER NECROSIS VIRUS

CNV is comprised of isometric particles about 31 nm in diameter that contain a monopartite ssRNA genome. It belongs to the tombusvirus group and is usually found only in greenhouse cucumbers, but experimentally can infect a wide range of plants. Symptoms consist of chlorotic spots with pinpoint necrotic centers, which usually fall out, leaving small circular holes of various sizes. Leaves may be severely deformed, occasionally showing dark green enations. Plants are stunted, and the mottled fruits fail to develop to full size. During summer months, symptoms tend to become very mild (Dias and McKeen 1972). No sources of resistance have been reported for this virus.

MELON NECROTIC SPOT VIRUS

MNSV particles are about 30 nm in diameter, also contain an ssRNA, and may belong to the carmovirus group. MNSV also occurs in greenhouse-grown cucumber, melon, and watermelon, causing significant reductions in yield. Experimentally, the virus can infect squash, *Lagenaria*, a number of wild cucurbit species, and also some legumes. Symptoms consist of necrotic spots of different sizes and shapes occurring on leaves and stems of melons, whereas in cucumber, symptoms are usually confined to foliage. Fruits from infected plants are usually small and may exhibit necrotic spots. A special strain has been reported to cause very severe leaf and stem necrosis, followed by death in watermelon (Avgelis 1989). Seed transmission of this virus in melon ranges from 1 to 15%. No infection occurs when melon seeds from infected plants are sown in soil free of the fungal vector, but when they are sown in soil containing the virus-free fungus, seedlings become infected. This type of seed transmission is called *vector-mediated seed transmission*. MNSV shares several characteristics with CNV, but the two viruses differ in their physical and chemical properties and are unrelated serologically (Kishi 1966; Gonsalez-Garza et al. 1979). A number of melon varieties were reported to be either free of local and systemic symptoms or to respond to the virus with localized infection by a strain from California (Gonsalez-Garza et al. 1979).

Cucurbit Viruses Transmitted by Leafhoppers

Only one cucurbit virus, beet curly top (BCTV), is transmitted by leafhoppers. It can cause considerable damage to cucurbit crops.

BEET CURLY TOP VIRUS

Beet curly top virus is a geminivirus with isometric particles about 20 nm in diameter, which usually occur in pairs (geminate) and contain ssDNA. This virus is found mostly in the arid and semi-arid climates, where it can infect a large number of plant species. The leafhoppers *Circulifer tenellus* and *C. opacipennis* are the major vectors. Virus transmission can occur after a brief acquisition time. The minimum latent period in the vector before transmission is less than 4 hours. Infectivity lasts for the life of the vector, but will decrease with its age. BCTV can infect a large number of plant species, and many strains and variants have been reported. The virus is confined to the phloem and is transmissible mechanically only by special procedures, via a pin-prick or a high pressure method. Infected cucurbit plants are generally severely stunted, showing upward rolling of leaf laminae and rosetting of apical growth. Fruits are severely malformed (Giddings 1948; Thomas and Mink 1979). No information is available regarding sources of resistance in any of the cultivated cucurbits.

Cucurbit Viruses Transmitted by Nematodes

Nematode-transmitted viruses of cucurbits include two nepoviruses, tobacco ringspot virus (ToRSV) and tomato ringspot virus (TmRSV), which are spread by adults and larvae of the nematodes *Xiphinema americanum*, *X. rivesi*, and other related species, and also by *Longidorus* species. Viruses are acquired within 24 hours and are transmitted and retained for several weeks or months. These viruses do not multiply in the vectors, nor are they transmitted through eggs. Virus particles are associated with specific sites of the alimentary tracts, such as the stylet lumen or guiding sheath, or esophagus. ToRSV and TmRSV are mainly found in fields that have been cultivated after a prolonged period of rest. Undisturbed soil usually allows nematodes to multiply abundantly and acquire the viruses from infected weed species, then spread them to other plants. After a few years of intense cultivation, populations of these nematodes rapidly decrease (Stace-Smith 1984, 1985).

TOBACCO RINGSPOT VIRUS

ToRSV has particles about 28 nm in diameter with a bipartite ssRNA genome (RNA-1 and RNA-2), both of which are essential for replication and pathogenicity. Serologically, ToRSV is not related to TmRSV. It is easily transmitted mechanically and in seed and pollen of some melons (about 3%). ToRSV can infect a very large number of cultivated and wild plant species, including woody and herbaceous types. On cucurbits, symptoms consist of chlorotic spots formed by concentric rings; some varieties also develop foliar necrosis and may die prematurely. Many plants tend to recover from the acute stage of infection, however, and during the chronic stage, symptoms may appear to be mild, although fruit production is significantly affected. Cucumbers, melons, and watermelons are particularly susceptible to this virus (Stace-Smith 1985). No resistance has been reported in cucumber, melon, or watermelon. Resistance was found in several cultivated and wild squash (*Cucurbita* spp.) and some *Cucumis* spp. (Provvidenti et al. 1978).

TOMATO RINGSPOT VIRUS

TmRSV shares many characteristics with tobacco ringspot virus: particles are about 28 nm in diameter with a bipartite RNA genome. Its host range comprises hundreds of plant species, both woody and herbaceous. It is easily transmitted mechanically and in seeds of some plant species, but not in cucurbits. Symptoms resemble those incited by ToRSV and consist of chlorotic and necrotic mottle with ring-like spots, veinal netting, leaf distortion, and stem and apical necrosis. Infected plants usually survive the first stage of infection (acute stage) and may partially recover from severe symptoms to enter a chronic stage of infection. Fruits of infected plants may develop gray or brown corky spots. Particularly affected are summer and winter squash (*C. pepo*, *C. maxima*, *C. moschata*, and others), whereas in cucumber and melon, symptoms appear to be mild and often transitory. Nonetheless, plants are still infected by the virus, and consequently they cannot be reinfected by the same virus (Stace-Smith 1984). Resistance was located in some wild squash (*Cucurbita cylindrata*, *C. digitata*, *C. ecuadorensis*, *C. gracilior*, *C. palmata*, *C. palmeri*, and *C. sororia*) (Provvidenti et al. 1978; Provvidenti 1990). Resistance was also found in some accessions of *Lagenaria siceraria* (PI 188809 from The Philippines and PI 271353 from India (Provvidenti 1981).

Cucurbit Viruses Transmitted by Thrips

Tomato spotted wilt (TSWV) is the only virus transmitted by thrips that can infect some cucurbits.

TOMATO SPOTTED WILT VIRUS

Tomato spotted wilt virus is an RNA-containing virus with a tripartite genome encapsidated by particles about 70–90 nm. It is unique among plant viruses in that the virion is covered by a lipoprotein envelope. Based upon this characteristic, its distinctive size and morphology, and its status as the only virus transmitted in a circulative manner in thrip species, it is the only member of this plant viral group. Although highly unstable *in vitro*, it can be transmitted mechanically, but there is no evidence that it can be spread through infected seeds. The insect vectors belong to the family Thripidae, Order Tysanoptera, and include *Frankliniella fusca*, *F. occidentalis*, *F. schultzei*, *Scirtothrips dorsalis*, *Thrips setosus*, and *T. tabaci*. The vectors acquire viruses only during the larval stages and can transmit them both before and after pupation. Adults remain viruliferous throughout their life span, but the rate of transmission can vary.

Indirect evidence suggests that TSWV can replicate in thrips tissue. Most of the cucurbit species respond to this virus infection with a localized reaction. In recent years, however, it has been reported to infect systemically some species. A silver foliar mottle of watermelon in Japan was caused by a strain of this virus, which differs serologically from other strains (Ie 1970; Iwaki et al. 1984; Cho et al. 1989). Since this virus does not usually cause appreciable damage to cucurbits, no search for sources of resistance has been undertaken.

Cucurbit Viruses Transmitted by Whiteflies

Viral diseases transmitted by whiteflies are increasing in importance in tropical and other warm regions where the vectors are spreading rapidly due to increasing resistance to insecticides and the development of new biotypes. This group includes: beet pseudo-yellows (BPYV), cucumber vein yellowing (CVYV), cucumber yellows (CYV), lettuce infectious yellows (LIYV), melon leaf curl (MLCV), squash leaf curl (SLCV), and watermelon curly mottle (WCMoV).

BEET PSEUDO-YELLOWS VIRUS

BPYV is a geminivirus and is able to infect a wide range of plant species, including cucurbits. BPYV is transmitted by the common greenhouse whitefly, *Trialeurodes vaporarium*, which acquires it in a feeding period of one hour and retains it for six or more days. Leaves of affected plants develop irregular necrotic areas and eventually will die prematurely. Infected plants are usually very stunted (Duffus 1965). Sources of resistance were found in some wild cucurbit species, including *Cucumis anguria* var. *longipes*, *C. myriocarpus*, *C. zeheri*, and in the wild watermelon *Citrullus colocynthis* (Esteva et al. 1988). A certain tolerance was reported in some landraces of *C. melo* from Spain and from the Orient. An accession of *C. melo* var. *agrestis* displayed very mild symptoms under severe field infection in Spain (Esteva et al. 1989).

CUCUMBER VEIN YELLOWING VIRUS

A rod-shaped virus, CVYV has long particles of about 740–800 nm, containing a double-stranded DNA. It is therefore unique among plant viruses, because the other DNA-containing plant viruses exhibit spherical symmetry (Paris et al. 1988). CVYV is transmitted by *Bemisia tabaci* and also mechanically. Experimentally, however, both methods of transmission appear to be inefficient. CVYV can infect many cucurbit species. Symptoms range from chlorotic or necrotic dots to severe vein yellowing and stunting. Originally it was found in Israel and named bottlegourd mosaic virus (Cohen and Nitzany 1960; Sela et al. 1980). No sources of resistance are known.

CUCUMBER YELLOWS VIRUS

Cucumber yellows virus is probably a closterovirus, as it has long flexuous particles about 1000 nm, and is mostly confined to the phloem. It is transmitted by the whitefly *Trialeurodes vaporarium*, but not mechanically (Purcifull et al. 1984). It was originally reported from Japan (Purcifull et al. 1984), where it infected cucumber and melon. CYV was also found to occur in melon in France (Hibi and Furuki 1985). Symptoms consist of interveinal chlorotic spots, which eventually enlarge and become golden yellow, whereas veins remain green. There is no major reduction in size of the main stem, but the number of lateral shoots is drastically reduced. The host range seems to be confined to Cucurbitaceae (Yamashita et al. 1979; Lot et al. 1982). No information is available regarding resistance.

LETTUCE INFECTIOUS YELLOW VIRUS

LIYV belongs to the closterovirus group, hence virus particles are long, filamentous, and rod shaped, about 1200–2000 nm, and contain a monopartite ssRNA genome. Serologically, it is not related to any of the other members of this group. The major vector is *Bemisia tabaci*, which can acquire the virus after several minutes of feeding and retain it for a few days. LIYV is mechanically transmissible, but there is no indication that it is seedborne. It is very destructive in lettuce, and it can infect several cucurbits, including cucumber, melon, squash, and watermelon. Symptoms consist of interveinal yellowing of the leaf lamina, green mosaic, pronounced leaf curling of the young leaves, and leathery consistency of older leaves. Infected plants are usually stunted with a poor fruit set or incomplete fruit development (Duffus et al. 1986). No sources of resistance have been reported for cucurbit species.

MELON LEAF CURL VIRUS

Melon leaf curl virus is a geminivirus, with geminate particles about 22 x 38 nm containing a ssDNA genome. Serologically it differs from squash leaf curl virus (SLCV). It is spread in nature by *Bemisia tabaci*, and affects melon, cucumber, squash, watermelon, and bean (*Phaseolus vulgaris*). MLCV is mechanically transmissible, but no information is available regarding whether it is seedborne. Symptoms caused by this virus are similar to those incited by SLCV, and include reduction of leaf size, thickening of veins, upward curling, enations, and plant stunting (Duffus et al. 1985). No sources of resistance have been reported.

SQUASH LEAF CURL VIRUS

Squash leaf curl virus is a recently recognized geminivirus (1981), with particles about 22 × 38 nm containing a single strand of RNA. Serologically, it is related to bean golden mosaic virus and cassava latent virus. In nature this virus is spread by the whitefly *Bemisia tabaci*, in which it is circulative with a relatively long latent period. SLCV can be transmitted mechanically, provided that plants are very young and a phosphate buffer (pH 8.5 or 8.8) is used. In squash, it causes reduction of leaf size, thickening of veins, upward curling, enations, and plant stunting. SLCV has been found in Southern California and Mexico, where it causes severe economic losses (Flock and Mayhew 1981; Cohen et al. 1983). Tolerance was reported in some varieties of *C. moschata* and *C. pepo* (McCreight 1984), but whether it can be exploited remains to be seen. Apparently, this virus is not able to infect cucumber, melon, or watermelon (Cohen et al. 1983).

WATERMELON CURLY MOTTLE VIRUS

This geminivirus (particles 20 × 30 nm) is serologically related to bean golden mosaic and spread in nature by *Bemisia tabaci*. The relationship between this virus and SLCV remains inconclusive. It can be transmitted mechanically, and its host range extends beyond those species reported for SLCV. WCMV infects mostly cucurbits and legumes. Originally found in Arizona, it is also widespread in the desert areas of the southwestern US and Mexico. On watermelon, as well as other cucurbits, symptoms consist of a prominent leaf curling, rugosity, mottling, vein-banding, and plant stunting (Brown and Nelson 1986, 1989). No sources of resistance have been reported for this virus.

Viruses With Unknown Biological Vectors

Viruses and one viroid whose biological vectors have not been definitively established include cucumber green mottle mosaic virus (CGMMV), cucumber leaf-spot virus (CLSV), cucumber pale fruit viroid (CPFV), and Ournia melon virus (OMV).

CUCUMBER GREEN MOTTLE MOSAIC VIRUS

CGMMV is a tobamovirus with rod-shaped particles about 300 nm, containing a single strand of RNA. It is easily transmitted by foliage contact and soil contamination and through seed. No biological vector is known. CGMMV has a restricted host range, involving mainly cultivated cucurbits. Several strains have been reported that can cause a variety of symptoms, ranging from a mild foliar mottle to a very prominent bright yellow mosaic. Leaves and fruits may be distorted and reduced in size. The watermelon strain can induce internal discoloration and decomposition of fruits. This virus has been reported from Europe and Asia, particularly in greenhouse grown plants (Hollings et al. 1975).

No high level of resistance has been located in cucumber and related species. Several Asian cucumbers, although infected, remain relatively symptomless. These lines cannot be considered tolerant, because the reduction in yield is similar to that of known susceptible varieties. Several wild *Cucumis* species were proven symptomless carriers of the virus, including *C. africanus*, *C. ficifolius*, *C. figarei*, *C. meeusii*, and *C. myriocarpus*. *Cucumis zeyheri* was considered to be immune (Rajamony et al. 1987). In another study, in crosses between the CGMMV-resistant *C. anguria* with a susceptible line of *C. myriocarpus*, it was established that resistance is conferred by the monogenically dominant factor *Cgm* (Nijs 1982). Resistance in a muskmelon line was demonstrated to be polygenic recessive (Rajamony et al. 1990).

CUCUMBER LEAF-SPOT VIRUS

Cucumber leaf-spot virus is a tombusvirus with isometric particles about 28 nm long, containing ssRNA. It occurs mostly in Europe and the Middle East in cucumbers, in which it incites irregularly shaped light green to yellowish spots with necrotic centers. Affected plants are usually stunted, and fruits may develop longitudinal chlorotic streaks, which eventually will turn necrotic. No biological vector has been identified for this virus, but it is transmitted mechanically and via seeds (about 1%). Experimentally, this virus was able to infect herbaceous hosts of 26 species, which responded with a localized infection (Weber 1986). No sources of resistance have been reported.

CUCUMBER PALE FRUIT VIROID

CPFV has been reported from The Netherlands (Dorst and Peters 1974) and Japan (Sano et al. 1981). Viroids differ from viruses in that the ssDNA genome is not encapsidated and codes for no known gene products. Comparative studies have shown that CPFV is identical to hop stunt viroid in its biological properties (Nwauzo and Brown 1975). It has been found in cucumber grown under greenhouse conditions. The main characteristics of the disease are pale green fruits and crumpled flowers, with foliar rugosity and chlorosis. In melon, watermelon, and waxgourd, the disease appears to be more severe. These plants are stunted, have a bushy appearance, and eventually die prematurely. CPFV is more common in spring plantings, and can be transmitted with sap during pruning and grafting. The host range includes many cucurbit species (Sano et al. 1981). It was reported that one Japanese watermelon and a variety of *Cucurbita moschata* were not infected by this viroid (Dorst and Peters 1974). A *Momordica* species was also not infected by CPFV (Sano et al. 1981).

OURNIA MELON VIRUS

Ournia melon virus is a newly recognized plant virus belonging to a new group. Particles are rather stable and resemble those of both geminiviruses and alfalfa mosaic virus. They are 30 to 37 nm long and 18.5 wide, containing ssRNA. Both ends of these particles are sharply triangular in profile, and hence are unique among plant virus particles. OMV was originally isolated from melons grown in the

Ournia District in Northern Iran, from which it takes its name. No biological vector is known, but it can be easily transmitted mechanically. OMV has a wide host range. Among cucurbits, it can infect cucumber, melon, *Luffa acutangula*, and *Cucumis metuliferus*. Symptoms include chlorotic ringspots, mosaic, leaf puckering, and plant stunting (Lisa et al. 1988). Very little is known about resistance in cucurbit species, but with experimental inoculations, *Cucurbita pepo* and *C. maxima* remained free of symptoms (Lisa et al. 1988).

Conclusions and Future Perspectives

Significant progress has been made toward the identification and characterization of the viruses affecting cucurbits in different regions of the world and in the search for genetic sources of resistance. In contrast to sources of resistance to fungal and bacterial pathogens, genes conferring resistance to viral infections have been proven rather stable, although stable resistance may be narrow in effect, and therefore specific to particular viral strains. Thus, a gene may confer resistance only to a specific strain or pathotype of a given virus, necessitating an accumulation or pyramiding of genes for the control of all known pathotypes. Private and state institutions are developing a new generation of highly productive cucurbits that are resistant to viral as well as fungal and bacterial diseases. Crops grown under protected conditions will be the major beneficiaries, but sanitation and vector control should be strictly enforced to maximize the benefit deriving from horticulturally advanced varieties. Local resistance-breeding programs should receive encouragement and adequate financial support, as varieties adapted to one set of production conditions may not be suited to others in which the same viral diseases are problematic.

For cucumber, breeders have at their disposal genes controlling resistance to all major aphid-transmitted viruses. Several American, Chinese, European, and Japanese seed companies have incorporated genes for resistance to viruses, fungi, and bacteria. All these newly available varieties should be evaluated in regions where viral diseases are a chronic problem for yield, quality, disease resistance, and acceptability by local markets. A major effort should be made to determine which of the whitefly-transmitted viruses are present in each tropical and subtropical region and to search for sources of resistance. In recent years, these viruses have assumed great economic importance.

For melon, resistance genes are available for CMV, PRSV, WMV, and ZYMV, as well as for several fungal and bacterial diseases. Presently, a few commercial varieties possess resistance to some viruses, but private and state institutions are working toward multi-resistant varieties. When available, they should be evaluated under different conditions. For example, available evidence indicates that the gene *Zym* for resistance to ZYMV in melon is not functional against some strains of this virus. A search for a more effective gene is under way. Meanwhile, a solution of the problem rests in protecting melon plants with a mild strain of the virus either by conventional or transgenic means. Some of the whitefly-transmitted viruses can cause considerable losses in this crop.

For summer squash, progress toward a solution of viral problems has been slow. No very effective sources of resistance have been located in *Cucurbita pepo* and its closely related species *C. texana* and *C. fraterna*. Thus, resistance has been derived from other cultivated and wild species. Of particular value are *C. moschata* 'Nigerian Local' and *C. ecuadorensis*, which are multiresistant. Nonetheless, genetic incompatibility among species remains a formidable barrier that has retarded progress. Several institutions are presently involved in breeding summer and winter squash with multiresistance to viruses by using conventional and laboratory-based methods. In a few years, some varieties will be available, but again, local breeding is necessary, particularly for the varieties that are preferred by local markets.

For watermelon, resistance to ZYMV is available from two sources: *C. lanatus* and *C. colocynthis*. No high level of resistance has been found for PRSV-W, but field tolerance to this virus appears to be available in ZYMV-resistant breeding lines. A good level of resistance to WMV is also available in some wild species (*C. colocynthis*).

Although excellent results have been obtained in breeding for resistance to major cucurbit viruses, efforts must also be made to locate sources of resistance for viruses considered of minor importance because they are present only in particular areas or regions of the world. Sources of resistance are also urgently needed for a number of cucurbits, including species of *Acanthosicyos*, *Benincasa*, *Coccinia*, *Corallocarpus*, *Cucumeropsis*, *Cyclanthera*, *Luffa*, *Hodgsonia*, *Momordica*, *Praecitrullus*, *Sechium*, *Telfairia*, *Trichosanthes*, and others that have food value.

With concentrated effort, results can be rapid and rewarding. Hence, systematic surveys should be made in the major cucurbit-growing areas of the world to monitor the occurrence of new viral diseases. Because of increased knowledge of plant viruses and more effective international cooperation among researchers, it is now possible to accomplish more in less time. As mentioned above, considering the economic importance of some whitefly-transmitted viruses in the tropics, a special effort must be made to locate sources of resistance to these viruses as soon as possible.

Not all the cucurbit collections available in the various countries of the world have been thoroughly evaluated for resistance to viruses. Thus, it is possible that a large range of genetic diversity remains untapped. More collections are needed at the center of origin of each species, which may reveal better and more versatile germplasm. Most of the genes for resistance to viral diseases of cucumber and melon have derived from landraces collected in the Orient, whereas resistance genes in squash and watermelon originated from Central American and African germplasm. Consequently, every effort must be made to maintain and preserve landraces and wild and semiwild relatives of cultivated species in their wild habitats, where they may continue to evolve.

As traditional breeding methods are augmented by additional techniques, such as early embryo rescue, tissue culture, protoplast regeneration, and cellular fusion, more benefits are forecast by advances in biotechnology. Presently, efforts are directed toward transgenic cross protection or symptom interference based upon the insertion of virus coat proteins into the cucumber, melon, and squash genomes. These transgenic lines are receiving intense evaluation. Still, these highly glamorized, new methods of viral resistance clearly have their limitations, e.g. strain-specificity, virus titer dependence, the potential for encapsidation and dissemination of unrelated viruses, and appearance of new strains able to overcome resistance. More significant progress can be achieved by transferring naturally occurring resistance genes from one species to another. A number of approaches are being taken to clone them in the absence of defined biochemical products.

One of the more promising methods may be mapping resistance genes relative to polymorphic markers. Using restriction fragment length polymorphism (RFLP) for chromosome walking, genes can be identified and eventually cloned *in vitro*. The construction of polymorphic marker maps are being facilitated by the adoption of techniques based on the polymerase chain reaction (PCR). The actual transfer of these genes is accomplished via biological vectors such as *Agrobacterium tumefaciens* or biolistic approaches. The benefit of cloning resistance and transferring of naturally occurring resistance genes is likely to be considerable, because these genes are the product of many thousands of years of evolution. In the foreseeable future, resistance genes may be manipulated *in vitro* in order to control new pathotypes. Eventually, these techniques will provide valuable information regarding the evolution of resistance in the everlasting struggle between plants and pathogens.

Finally, genes for resistance provide excellent tools for virus identification. Laboratories in developed countries are usually well equipped, thus making virus characterization a relatively simple task. Since facilities are very limited in most other countries, very specific tests can be accomplished with host plants possessing a single gene for resistance. These tests are perhaps the most useful of those based upon biological response and, in most cases, they offer the only opportunity to identify viral pathotypes. Resistant plants are also useful in separating individual viruses from mixtures.

Literature Cited

Anno-Nyako, F. O. 1988. Seed transmission of Telfairia mosaic virus in fluted pumpkin (*Telfairia occidentalis* Hook) in Nigeria. J. Phytopathol. 12:85–87.

Atiri, G. I., and A. Varma. 1983. Development of improved lines of *Telfairia occidentalis* Hook resistant to mosaic disease. Trop. Agr. (Trinidad) 60:95–96.

Avgelis, A. D. 1989. Watermelon necrosis caused by a strain of melon necrotic spot virus. Plant Path. 38:618–622.

Brown, J. K., and M. R. Nelson. 1986. Whitefly-borne viruses of melon and lettuce in Arizona. Phytopathology 76:236–239.

Brown, J. K., and M. R. Nelson. 1989. Characterization of watermelon curly mottle virus, a geminivirus distinct from squash leaf curl virus. Ann. Appl. Biol. 115:243–252.

Bhargava, B. 1977. Some properties of two strains of watermelon mosaic virus. Phytopath. Z. 88:199–208.

Campbell, R. N. 1971. Squash mosaic virus. Descriptions of Plant Viruses No. 43. Kew, Surrey, England: Commonw. Mycol. Inst./Assn. Appl. Biol.

Cho, J. J., R. F. L. Mau, T. L. German, R. W. Hartmann, S. L. Yudin, D. Gonsalves, and R. Provvidenti. 1989. A multidisciplinary approach to management of tomato spotted wilt virus in Hawaii. Plant Dis. 73:375–383.

Cohen, S., E. Gertman, and N. Kedar. 1971. Inheritance of resistance to melon mosaic virus in cucumbers. Phytopathology 61:253–255.

Cohen, S., and F. E. Nitzany. 1960. A whitefly-transmitted virus of cucurbits in Israel. Phytopath. Mediter. 1:44–46.

Cohen, S., J. E. Duffus, R. C. Larsen, H. Y. Liu, and R. A. Flock. 1983. Squash leaf curl virus: Purification, serology, and vector relationships of the whitefly transmitted geminivirus. Phytopathology 73:1669–1673.

Dias, H. F., and C. D. McKeen. 1972. Cucumber necrosis virus. Descriptions of Plant Viruses No. 82. Kew, Surrey, England: Commonw. Mycol. Inst./Assn. Appl. Biol.

Dorst, H. H. M. van, and D. Peters. 1974. Some biological observations on pale fruit, a viroid-incited disease of cucumber. Neth. J. Plant Pathol. 80:85–96.

Duffus, J. E. 1965. Beet pseudo-yellows virus, transmitted by the greenhouse whitefly (*Trialeurodes vaporarium*). Phytopathology 55:450–453.

Duffus, J. E., H. Y. Liu, and M. R. Johns. 1985. Melon leaf curl virus: A new geminivirus with host and serological variations from squash leaf curl virus. Phytopathology 75:1312.

Duffus, J. E., R. C. Larsen, and H. Y. Liu. 1986. Lettuce infectious yellows virus—A new type of whitefly-transmitted virus. Phytopathology 76:97–100.

Enzie, W. D. 1943. A source of muskmelon mosaic resistance found in the Oriental pickling melon, *Cucumis melo* var. *conomon*. Proc. Amer. Soc. Hort. Sci. 43:195–198.

Esteva, J., F. Nuez, and J. Cuartero. 1988. Resistance to yellowing disease in wild relatives of muskmelon. Cucurbit Genet. Coop. Rep. 11:52–53.

Esteva, J., F. Nuez, and M. L. Gomez-Guillamon. 1989. Resistance to yellowing disease in muskmelon. Cucurbit Genet. Coop. Rep. 12:44–45.

Fisher, H. U., and B. E. L. Lockhart. 1974. Serious losses in cucurbits caused by a watermelon mosaic virus in Morocco. Plant Dis. Rep. 58:143–146.

Flock, R. A., and D. Mayhew. 1981. Squash leaf curl, a new disease of cucurbits in California. Plant Dis. 65:75–76.

Francki, R. I. B., D. W. Mossop, and T. Hatta. 1979. Cucumber mosaic virus. Descriptions of Plant Viruses No. 213. Kew, Surrey, England: Commonw. Mycol. Inst./Assn. Appl. Biol.

Freitag, J. H., and K. S. Milne. 1970. Host range, aphid transmission, and properties of muskmelon vein necrosis virus. Phytopathology 60:166–170.

Giddings, N. J. 1948. Curly top of muskmelon. Phytopathology 38:934–936.

Gonsalez-Garza, R., D. J. Gumpf, A. N. Kishaba, and G. W. Bohn. 1979. Identification, seed transmission, and host range pathogenicity of a California isolate of melon necrotic spot. Phytopathology 69:340–345.

Greber, R. S. 1978. Watermelon mosaic viruses 1 and 2 in Queensland cucurbit crop. Austral. J. Agr. Sci. 29:1235–1245.

Greber, R. S., G. D. Perseley, and M. E. Herrington. 1988. Some characteristics of Australian isolates of zucchini yellow mosaic virus. Austral. J. Agr. Res. 39:1085–1094.

Grumet, R. 1990. Genetically engineered plant virus resistance. HortScience 25:508–513.

Hollings, M., Y. Komuro, and H. Tochihara. 1975. Cucumber green mottle mosaic virus. Descriptions of Plant Viruses No. 154. Kew, Surrey, England: Commonw. Mycol. Inst./Assn. Appl. Biol.

Hibi, T., and I. Furuki. 1985. Melon necrotic spot virus. Descriptions of Plant Viruses No. 302. Kew, Surrey, England: Commonw. Mycol. Inst./Assn. Appl. Biol.

Hseu, S. H., H. L. Wang, and C. H. Huang. 1985. Identification of a zucchini yellow mosaic virus from *Cucumis sativus*. J. Agr. Res. China 34:87–95.

Huang, C. H., Y. J. Chao, C. A. Chang, S. H. Hseu, and C. M. Hsiao. 1987. Identification and comparison of different viruses on symptom expression in loofah. J. Agr. Res. China 36: 414–420.

Ie, T. S. 1970. Tomato spotted wilt virus. Descriptions of Plant Viruses No. 39. Kew, Surrey, England: Commonw. Mycol. Inst./Assn. Appl. Biol.

Iwaki, M., Y. Honda, K. Hananda, H. Tochihara, K. Hokama, and T. Yokoyama. 1984. Silver mottle disease of watermelon caused by tomato spotted wilt virus. Plant Dis. 68:1006–1008.

Jones, P., S. B. Angood, and J. M. Carpenter. 1986. Melon rugose mosaic virus, the cause of a disease of watermelon and sweet melon. Ann. Appl. Biol. 108:303–307.

Karchi, Z., S. Cohen, and A. Govers. 1975. Inheritance of resistance to cucumber mosaic virus in melons. Phytopathology 65:479–48.

Kishi, K. 1966. Necrotic spot of melon, a new virus disease. Ann. Phytopath. Soc. Japan 32:138–144.

Kooistra, E. 1969. The inheritance of resistance to Cucumis virus l in cucumber (*Cucumis sativus* L). Euphytica 18:326–332.

Lecoq, H., and M. Pitrat. 1984. Strains of zucchini yellow mosaic virus in muskmelon (*Cucumis melo* L). Phytopath. Z. 111:165–173.

Lisa, V., and G. Dellavalle. 1981. Characterization of two potyviruses from zucchini squash. Phytopath. Z. 100:279–286.

Lisa, V., and H. Lecoq H. 1984. Zucchini yellow mosaic virus. Descriptions of Plant Viruses No. 282. Kew, Surrey, England: Commonw. Mycol. Inst./Assn. Appl. Biol.

Lisa, V., R. G. Milne, G. P. Accotto, G. Boccardo, and P. Cacciagli. 1988. Ournia melon virus, a virus from Iran with novel properties. Ann. Appl. Biol. 112:291–302.

Lockhart, B. E. L. and H. U. Fisher. 1979. Host range and some properties of Bryonia mottle virus, a new member of the potyvirus group. Phytopath. Z. 96:244–250.

Lot, H., B. Dellecolle, and H. Lecoq. 1982. A whitefly transmitted virus causing muskmelon yellows in France. Acta Hort. 127:175–182.

Martin, M. W. 1960. Inheritance and nature of cucumber mosaic virus resistance in squash. Dissert. Abstr. 20:3462.

Meer, F. W. van der, and H. M. Garnett. 1987. Purification and identification of a South African isolate of watermelon mosaic virus, Morocco. J. Phytopath. 120:255–270.

McCreight, J. D. 1984. Tolerance in *Cucurbita* spp. to squash leaf curl. Cucurbit Genet. Coop. Rep. 7:71–72.

Munger, H. M. 1976. *Cucurbita martinezii* as source of resistance. Veg. Improv. Newsl. 18:4.

Munger, H. M., and R. Provvidenti. 1987. Inheritance of resistance to zucchini yellow mosaic virus in *Cucurbita moschata*. Cucurbit Genet. Coop. Rep. 10:80–81.

Nelson, M. R., and H. K. Knuhtsen. 1973. Squash mosaic virus variability: Review and serological comparison of six biotypes. Phytopathology 63:920–926.

Nijs, A. P. M. den. l982. Inheritance of resistance to cucumber green mottle mosaic virus (*Cgm*) in *Cucumis anguria* L. Cucurbit Genet. Coop. Rep. 5:57–58.

Nwauzo, E. E., and W. M. Brown. 1975. Telfairia (Cucurbitaceae) mosaic virus in Nigeria. Plant Dis. Rep. 59:430–432.

Paris, H. S., S. Cohen, Y. Burger, and R. Yodrph. 1988. Single gene resistance to zucchini yellow mosaic virus in *Cucurbita moschata*. Euphytica 37:27–29.

Pitrat, M. 1990. Gene list for *Cucumis melo* L. Cucurbit Genet. Coop. Rep. 13:58–70.

Pitrat, M., and H. Lecoq. 1983. Two alleles for watermelon mosaic virus l resistance in melon. Cucurbit Genet. Coop. Rep. 6:52–53.

Pitrat, M., and H. Lecoq. 1984. Inheritance of zucchini yellow mosaic virus resistance in *Cucumis melo* L. Euphytica 33:57–61.

Porter, R. H. 1928. Further evidence of resistance to cucumber mosaic in the Chinese cucumber. Phytopathology 18:143.

Provvidenti, R. 1977. Evaluation of vegetable introductions from the People's Republic of China for resistance to viral diseases. Plant Dis. Rep. 61:851–855.

Provvidenti, R. 1981. Sources of resistance to viruses in *Lagenaria siceraria*. Cucurbit Genet. Coop. Rep. 4:38–40.

Provvidenti, R. 1982. Sources of resistance and tolerance to viruses in accessions of *Cucurbita maxima*. Cucurbit Genet. Coop. Rep. 5:46–47.

Provvidenti, R. 1985. Sources of resistance to viruses in two accessions of *Cucumis sativus*. Cucurbit Genet. Coop. Rep. 8:12.

Provvidenti, R. 1986a. Reactions of accessions of *Citrullus colocynthis* to zucchini yellow mosaic virus and other viruses. Cucurbit Genet. Coop. Rep. 9:82–83.

Provvidenti, R. 1986b. Viral diseases of cucurbits and sources of resistance. Tech. Bull. No. 93. Taipei, Taiwan: Food and Fertiliz. Technol. Center. 16 pp.

Provvidenti, R. 1987. Inheritance of resistance to a strain of zucchini yellow mosaic virus in cucumber. HortScience 22:102–103.

Provvidenti, R. 1990. Viral diseases and genetic sources of resistance in *Cucurbita* species. In Biology and Utilization of the Cucurbitaceae. Ed. D. M. Bates, R. W. Robinson, C. Jeffrey. Ithaca, NY: Cornell University Press.

Provvidenti, R. 1991. Inheritance of resistance to the Florida strain of zucchini yellow mosaic virus in watermelon. HortScience 26:407–408.

Provvidenti, R., and R. W. Robinson. 1977. Inheritance of resistance to watermelon mosaic virus 1 in *Cucumis metuliferus*. J. Heredity 68:56–57.

Provvidenti, R., and R. W. Robinson. 1987. Lack of seed transmission in squash and melon plants infected with zucchini yellow mosaic virus. Cucurbit Genet. Coop. Rep. 10:81–82.

Provvidenti, R., and J. K. Uyemoto. 1973. Chlorotic leaf spotting of yellow summer squash caused by the severe strain of bean yellow mosaic virus. Plant Dis. Rep. 57:280–282.

Provvidenti, R., R. W. Robinson, and H. M. Munger. 1978. Resistance in feral species to six viruses infecting *Cucurbita*. Plant Dis. Rep. 62: 326–329.

Provvidenti, R., D. Gonsalves, and H. S. Humaydan. 1984. Occurrence of zucchini yellow mosaic virus in cucurbits from Connecticut, New York, Florida, and California. Plant Dis. 68:443–446.

Purcifull, D., and D. Gonsalves. 1984. Papaya ringspot virus. Descriptions of Plant Viruses No. 292. Kew, Surrey, England: Commonw. Mycol. Inst./Assn. Appl. Biol.

Purcifull, D. E., and E. Hiebert. 1979. Serological distinction of watermelon mosaic virus isolates. Phytopathology 69:112–116.

Purcifull, D., H. Hiebert, and J. Edwardson. 1984. Watermelon mosaic virus 2. Descriptions of Plant Viruses No. 293. Kew, Surrey, England: Commonw. Mycol. Inst./Assn. Appl. Biol.

Rajamony, L., T. A. More, and V. S. Sheshadri. 1990. Inheritance of resistance to cucumber green mottle mosaic virus in muskmelon (*Cucumis melo* L.). Euphytica 47:93–97.

Rajamony, L., T. A. More, and V. S. Sheshadri. 1987. Resistance to green mottle mosaic virus (CGMMV) in muskmelon. Cucurbit Genet. Coop. Rep. 10:58–59.

Regenmortel, M. H. V. van. 1972. Wild cucumber mosaic virus. Descriptions of Plant Viruses No 105. Kew, Surrey, England: Commonw. Mycol. Inst./Assn. Biol.

Risser, G., M. Pitrat, and J. C. Rode. 1977. Etude de la resistance du melon (*Cucumis melo* L) au virus de la mosaique du concombre. Annales Amelior. Plantes 27:509–522.

Robinson, R. W., H. M. Munger, T. W. Whitaker, and G. W. Bohn. 1976. Genes of the Cucurbitaceae. HortScience 11:554–568.

Robinson, R. W., N. F. Weeden, and R. Provvidenti. 1988. Inheritance of resistance to zucchini yellow mosaic virus in the interspecific cross *Cucurbita maxima* × *C. ecuadorensis*. Cucurbit Genet. Coop. Rep. 11:74–75.

Sano, T., M. Sasaki, and E. Shikata. 1981. Comparative studies on hop stunt viroid, cucumber pale fruit viroid and tomato spindle tuber viroid. Ann. Phytopath. Soc. Japan 47:599–605.

Schrijnwerkers, C. C. F. M., N. Huijberts, and L. Bos. 1991. Zucchini yellow mosaic virus: Two outbreaks in The Netherlands and seed transmissibility. Neth. J. Plant Path. 97:187–191.

Sela, I., I. Assouline, E. Tanne, S. Cohen, and S. Marco. 1980. Isolation and characterization of a rod-shaped, whitefly-transmissible DNA-containing plant virus. Phytopathology 70:226–228.

Sharma, O. P., H. L. Khatri, R. D. Bansal, and H. S. Komal. 1984. A new strain of cucumber mosaic virus causing mosaic disease of muskmelon. Phytopath. Z. 109:332–340.

Stace-Smith, R. 1984. Tomato ringspot virus. Description of Plant Viruses No. 290. Kew, Surrey, England: Commonw. Mycol. Inst./Assn. Appl. Biol.

Stace-Smith, R. 1985. Tobacco ringspot virus. Description of Plant Viruses No. 309. Kew, Surrey, England: Commonw. Mycol. Inst./Assn. Appl. Biol.

Takada, K. 1979. Studies on the breeding of melon resistant to cucumber mosaic virus. III. Inheritance of resistance of melon to cucumber mosaic virus and other characteristics. Bull. Vegetable Ornamental Crops Res. Station. Japan, Series A, No. 5, pp. 71–80.

Takada K., K. Kanazawa, K. Takatuka, and T. Kameno. 1979. Studies on the breeding of melon resistance to cucumber mosaic virus. I. Difference in resistance among melon varieties and the regional differences in their distribution. Bull. Vegetable Ornamental Crops Res. Station. Japan, Series A, No. 5, pp. 1–22.

Thomas, P. E., and G. I. Mink. 1979. Beet curly top virus. Descriptions of Plant Viruses No. 210. Kew, Surrey, England: Commonw. Mycol. Inst./Assn. Appl. Biol.

Thottappilly, G., and R. Lastra. 1987. The occurrence, properties and affinities of *Telfairia* mosaic virus, a potyvirus prevalent in *Telfairia occidentalis* (Cucurbitaceae) in South Western Nigeria. J. Phytopath. 119:13–24.

Vovlas, C., E. Hiebert, and M. Russo. 1981. Zucchini yellow fleck virus, a new potyvirus of zucchini squash. Phytopath. Mediter. 20:123–128.

Wang, Y-J., R. Provvidenti, and R. W. Robinson. 1984. Inheritance of resistance to watermelon mosaic 1 virus in cucumber. HortScience 19:587–588.

Wang, Y-J., R. W. Robinson, and R. Provvidenti. 1987. Linkage relationships of watermelon mosaic virus 1 in cucumber. Cucurbit Genet. Coop. Rep. 6:92.

Washek, R. L., and H. M. Munger. 1983. Hybridization of *Cucurbita pepo* with disease resistant *Cucurbita* species. Cucurbit Genet. Coop. Rep. 6:92.

Webb, R. E. 1965. *Luffa acutangula* for separation and maintenance of watermelon mosaic virus 1 free of watermelon mosaic virus 2. Phytopathology 55:1379–1380.

Webb, R. E. 1977. Resistance to watermelon mosaic virus 2 in *Citrullus lanatus*. Proc. Amer. Phytopath. Soc. 4:220.

Webb, R. E. 1979. Inheritance of resistance to watermelon mosaic virus l in *Cucumis melo* L. HortScience 14:265–266.

Webb, R. E., and G. W. Bohn. 1962. Resistance to cucurbit viruses in *Cucumis melo* L. Phytopathology 52:1221.

Weber, I. 1986. Cucumber leaf spot virus. Descriptions of Plant Viruses No. 319. Kew, Surrey, England: Commonw. Mycol. Inst./Assn. Appl. Biol.

Whitaker, T. W., and G. N. Davis. 1962. Cucurbits. New York: Interscience Publishers, Inc.

Yamashita, S., Y. Doi, K. Yora, and M. Yoshino. 1979. Cucumber yellows virus: Its transmission by the greenhouse whitefly *Trialeurodes vaporarium* (Westwood), and the yellowing diseases of cucumber and muskmelon caused by this virus. Ann. Phytopath. Soc. Japan 45:484–496.

CHAPTER 2

Breeding for Viral Disease Resistance in Cucurbits

HENRY M. MUNGER

Introduction

Cucumbers (*Cucumis sativus* L.) grown in the United States in 1950 were susceptible to all known cucumber diseases; by 1989 virtually all were resistant to at least two and in some cases five or six diseases. Resistance to cucumber mosaic has been a key ingredient of the change. While there has been substantial progress in resistance to fungal diseases in melons and watermelons, viral diseases of these species remain to be conquered, and almost no disease resistance of any kind is available commercially to squash, for which there is a need, especially in *Cucurbita pepo* varieties. Resistance to all the important diseases is known in these crops, however, and this chapter focuses on lessons learned from the breeding of cucumbers, in hopes that these lessons may speed a similar transformation of the other cucurbits to multiple disease-resistant varieties.

Dr. Munger is in the Departments of Plant Breeding and Vegetable Crops, New York State College of Agriculture and Life Sciences, Cornell University, Ithaca, NY 14853. The author is pleased to recognize the many who contributed to the results reported here. College and department administrators provided the essential continuity of support through both State and Hatch Funds. A number of plant pathologists, especially Charles Chupp, Robert Wilkinson, Arden

It appears that resistance to viral diseases is the need of first priority with melons (*Cucumis melo* L.) and summer squash (*Cucurbita pepo* L.). In winter squash, where the predominant type is Butternut (*Cucurbita moschata* Poir.), fewer losses usually occur from disease than in the other species. Watermelon varieties with resistance to one or more fungal diseases are now in use, and viral diseases present the greatest threat to their production in the United States (J. M. Crall 1990, personal communication). Viral diseases have been devastating on watermelon in Egypt, a situation that may be representative of a considerable area around the Mediterranean.

Despite the title, resistance to some fungal diseases will be discussed because the increased longevity of mosaic-resistant cucumbers has revealed the importance of other diseases and made resistance to them essential to maximize the usefulness of mosaic resistance.

Resistance in this chapter means resistance to a disease such as cucumber mosaic, and not necessarily to the pathogen, cucumber mosaic virus (CMV). Breeders select for resistance to disease on the basis of reduced symptoms and better growth, seldom knowing whether this is brought about by some mechanism such as reduced replication or movement of the virus or by tolerance to the presence of the virus. It is convenient to speak of resistance to CMV or watermelon mosaic virus (WMV), but more accurate to talk about resistance to the disease.

Cucumber (*Cucumis sativus* L.)

The history of disease resistance in cucumbers in this country dates back to about 1927, when 'Chinese Long' was introduced from China

---

Sherf, and Thomas Zitter on Cornell's Ithaca campus and Rosario Provvidenti at Geneva, gave motivation, technical advice, and virus isolates that were indispensible. John Carew, Edwin Oyer, Philip Minges, and Oscar Pearson contributed to the setting of objectives and horticultural evaluation of breeding lines. Carl English as experimentalist during most of the period and Margaret Wilson in later years gave excellent support in field and greenhouse operations. The assistance of several graduate students besides those named was much appreciated.

by R. H. Porter of Iowa State University (Porter 1929). This variety has been the main source of mosaic resistance in cucumbers, but many years of work were needed to breed varieties fully satisfactory in other respects. Most of the mosaic resistant varieties in use today are traceable to the work started in 1938 at Cornell University by Oved Shifriss and C. H. Myers. My involvement with their material began in 1942. Shifriss et al. (1942) concluded that homozygosity of three partially dominant genes was needed for high resistance, thus our initial approach was to select and backcross the most resistant plants in F_2 generations. This approach was reasonably successful with pickles in that second backcross progenies given to the Heinz Co. soon resulted in the release of Ohio MR17, which in turn was the principal recurrent parent to which scab resistance was added in breeding Wisconsin SMR18 (Walker et al. 1953). Even though widely used, these varieties and their derivatives are not highly resistant to cucumber mosaic and appear to be almost susceptible under severe CMV stress.

We abandoned alternate crossing and selfing with slicing types when we found lower resistance after each backcross. What did pay off was the cross between two 1948 progenies derived from first or second backcrosses, which showed moderate mosaic resistance and different horticultural features (Figure 1). Transgressive segregates for resistance and the best combinations of fruit characteristics were selected through several generations of inbreeding until 1955. Row 55–610 was outstanding and after further trials was named 'Tablegreen'. The next year 56–388 was identified as earlier and more productive. After the addition of scab resistance, it became 'Marketmore'. The numbered lines appear as sources of CMV resistance in the pedigrees of some cucumbers bred elsewhere. Although both have a high level of mosaic resistance, in their first commercial trials it became apparent that the resulting improved longevity increased the danger of scab, caused by *Cladosporium cucumerinum*, previously considered a minor disease in New York. This result demonstrated that when one disease is a major limitation, assessing the importance of other diseases is difficult. It suggests that we may expect greater advances with melons when breeding for viral resistance if we take advantage of resistance to *Fusarium* and mildews already present in good varieties.

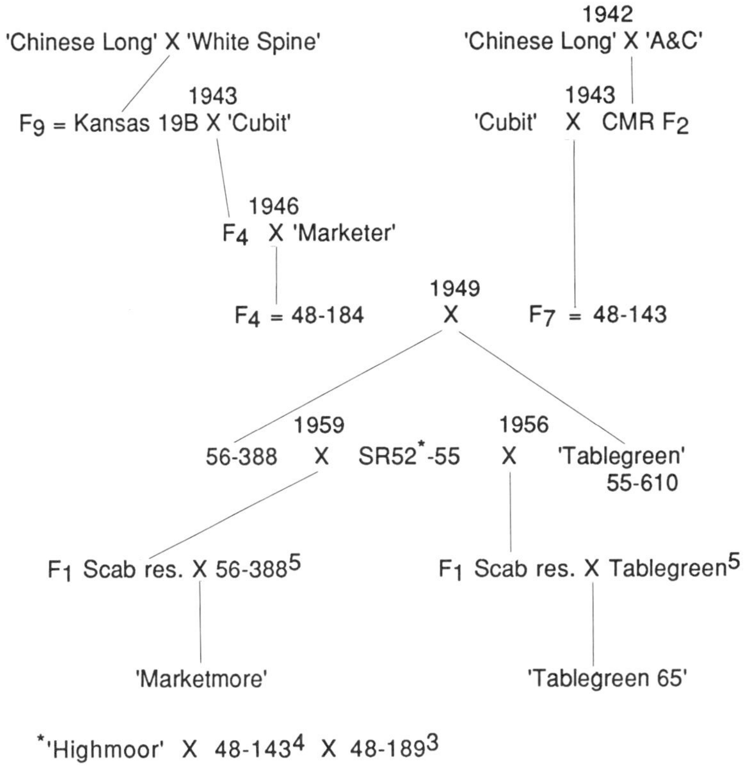

Figure 1. *Pedigrees of some cucumber mosaic resistant (CMR) cucumbers bred at Cornell University.*

The addition of scab resistance to the mosaic resistant slicing types was not as easy as one would expect from its inheritance as a single dominant gene, because it was associated with shorter fruit length. Solving this problem was important to make resistant varieties more useful but also led indirectly to our most important advance in speeding the breeding for viral disease resistance. This came about unexpectedly by a devious route that began with the outstanding work of W. C. Barnes at the Charleston station of Clemson University in South Carolina. Barnes bred the first multiple resistant cucumbers, of which the best slicer is 'Poinsett', with resistance to anthracnose, angular leafspot, and downy and powdery mildews

(Barnes 1961). His varieties have been extremely important in the southern United States and many tropical countries but have lacked the mosaic and scab resistance needed in cooler areas. Scab resistance was clearly needed as nearby as North Carolina. Barham and Winstead (1959) at North Carolina State University backcrossed the gene for scab resistance into two Clemson varieties, but the scab-resistant versions had distinctly shorter fruit and were not much used. Forewarned of this potential difficulty in breeding scab resistant versions of mosaic-resistant slicers, we made eight backcrosses to 'Tablegreen' and evaluated over 25 homozygous scab-resistant lines following the last backcross. Most of the lines had short fruit, but one with length equal to Tablegreen was named 'Tablegreen 65' (Wilson 1968).

A similar backcross procedure gave rise to 'Marketmore'. Released in 1965, it was quickly accepted in the northernmost states because of its combined mosaic and scab resistance, high yields, and high percentage of marketable fruit. However, it produced light-colored fruit when maturing at high temperatures, limiting its use even as far north as New Jersey and New York. This defect was remedied by backcrossing into it the recessive gene *u* for uniform fruit color from 'Tablegreen', which at that time was unique for color among U.S. varieties. This made the resulting 'Marketmore 70' much more widely adapted. Strong (1930) had reported on inheritance of the gene we now call "uniform," but its value was not recognized because he named its common dominant allele "mottled" and because the recessive he called "non-mottled" occurred in varieties with light green color. The *u* gene in mosaic-resistant varieties came from 'Chinese Long', in which its value was likewise unrecognized until recombined with darker background fruit color.

With the life of its vines extended by mosaic and scab resistance, the susceptibility of 'Marketmore' to mildew became increasingly apparent and was remedied by backcrossing into it two recessive genes for powdery mildew resistance (PMR) from C. E. Peterson's 'Spartan Salad'. The resulting version is 'Marketmore 76', with considerably widened geographic adaptation. The designation of this variety indicates its release 20 years after its desirable genetic background was first identified in row 56–388. The usefulness of the virus resistance of Marketmore was greatly enhanced by improvement in color and mildew resistance. Some reported usage of 'Marketmore 76' in southern states suggests that just as we overestimated the importance of mosaic as compared to other diseases in New York, we may have underestimated its importance farther south.

Continuous Backcrossing to Add Mosaic Resistance

With a mixture of motives, which at that time did not include virus resistance, we undertook in the mid-1960s to devote a small part of our effort on cucumbers to the improvement of 'Poinsett'. In part this was to move the adaptation of 'Poinsett' northward and in part to determine whether we had truly broken the linkage between scab resistance and short fruit. If linkage had been broken, transferring scab resistance from 'Tablegreen 65' to 'Poinsett' should lengthen rather than shorten the fruit. The *u* gene was added in a separate backcross program and brought together with scab resistance in 'Poinsett 76'. This indeed has slightly longer fruit than the original, which it has largely replaced. The relevance of this effort to virus resistance is that it caused seedsmen to request that we add mosaic resistance.

By then we were in the 1970s, thirty-five years after mosaic resistance work started at Cornell, and it had taken most of that time to breed a fully satisfactory slicing variety. To start again with the super-susceptible 'Poinsett' seemed unrealistic, but the limited success of our earlier use of the backcross method (i.e., alternating backcrossing and selfing) challenged us to try a new approach. Both the original work of Shifriss et al. (1942) and experience with many commercial hybrids indicated intermediate resistance in crosses between mosaic resistant × susceptible parents. It seemed that this might permit continuous backcrossing without alternate selfing and thereby speed the process of adding resistance. At the same time it might permit more rigorous elimination of plants not carrying resistance genes at all the important loci because every locus would have a gene for susceptibility from 'Poinsett'.

Following the initial cross of 'Tablegreen 65' × 'Poinsett', five backcrosses were made in three years. The numbers of progenies and plants selected per progeny are shown in Table 1. This series of backcrosses was followed by several generations of selfing and selection for resistance, which in the end gave a large number of lines with resistance close to that of 'Marketmore' and designated as 'Poinsett 83'. A substantial number of lines were eliminated because of late maturity, which was inadvertently retained from the 'Tablegreen 65' parent in selecting the most vigorous plants in the backcross F_1 progenies. The remaining lines were not bulked because they varied slightly in maturity, fruit length, and color, and possibly in genes for resistance, which might segregate if the lines were allowed to intercross. This work showed that resistance conditioned by several genes could be transferred more rapidly and retained at a higher level by continuous backcrossing than by alternate backcrossing and selfing.

Table 1. *Breeding 'Poinsett 83' by continuous backcrossing of cucumber mosaic resistance into 'Poinsett 76' (P76) with 'Tablegreen 68' as the source of resistance.*

	GENERATION	NUMBER OF PLANTS GROWN	NUMBER OF RESISTANT PLANTS used as parents for the next generation
Field 1974	F_1	6	
Greenhouse	BC_1	24	8
Field 1975[1]	BC_2	192	5
Greenhouse	BC_3	120	4
Field 1976[1]	BC_4	128	6
Field 1977[1]	BC_5 F_2	720	43
Field 1978[1]	BC_5 F_3	1376	42
Field 1979[1]	BC_5 F_4	1008	

[1]Each progeny in a field planting had either 6 or 8 hills with 2 plants per hill and 2 replicates. In the F_3 and F_4, one replicate was self-pollinated and one was picked at market stage to evaluate maturity.

Potyvirus Resistance in Cucumbers

'Poinsett 83' had resistances to cucumber mosaic and scab added to the original four resistances, but more remained to be done. Zucchini yellow mosaic virus (ZYMV) had been found in the United States in 1981 and soon had a widespread and devastating effect on certain cucurbit fields. With Provvidenti's finding (1985) of resistance to this and three other viruses in 'Tai Chung Mou Gua-1', we began adding resistance to ZYM, papaya ringspot (caused by PRSV, formerly WMV-1), and watermelon mosaic (caused by WMV, formerly WMV-2) to 'Poinsett 83'. This process is essentially complete and has been facilitated by the finding that selection for ZYM resistance seems to give resistance to WM as well. There may also be some association between ZYM and PR resistance, but the extent of this is not yet clear. Plants selected for high ZYM resistance in F_2 progenies usually give uniformly resistant F_3s, thereby confirming the recessiveness described by Provvidenti. Nevertheless, recessiveness cannot be complete because resistance has been retained through five backcrosses, some of which were made without alternate selfing, as F_1 plants with relatively low resistance were selected as parents.

Having assumed that the addition of mosaic and scab resistance to 'Poinsett' would adapt it to northern production, we were disappointed to find that it was defoliated prematurely by some leafspot in most of our trials. When we learned that 'Poinsett' is highly susceptible in the South to target leafspot (TLS) caused by *Corynespora cassiicola*, we assumed that this disease had occurred in our plots (Lane and Munger 1985). Since we initiated a backcross program to add *Corynespora* resistance, that pathogen has not been found in our fields but instead another fungus, *Ulocladium cucurbitae*, has been isolated consistently (Lane and Munger 1983, Zitter 1989). In spite of this finding, we continued to add *Corynespora* resistance and, having done so, were surprised to find that the new line had little or no leafspot. Because *Ulocladium* has always been the fungus isolated from these plots, it appears that one can obtain *Ulocladium* resistance by breeding for *Corynespora* resistance. The relevance of this experience is that we learned of the leafspot problem only because of breeding for viral resistance and found that the full value of virus resistance can be realized only if leafspot susceptibility is corrected.

We are combining the additional virus disease resistance with *Corynespora* resistance. When this is completed, six additional resistances will have been added to the original four in 'Poinsett'. A parallel program will culminate at about the same time with a 'Marketmore' with resistance to eight diseases, lacking only the angular leafspot and anthracnose resistance of 'Poinsett'. Cucumber breeding at Cornell began 50 years ago with the single objective of removing what was perceived as the main limitation to cucumber growing in New York, susceptibility to cucumber mosaic. It evolved into a program with multiple objectives as the removal of each limiting factor revealed the existence of others nearly as important.

Cucurbita Species

Squashes are one of the few vegetables with representatives of more than one species, each with some distinctive reactions to viruses, and the only cucurbit, with the possible exception of watermelon, for which it has been necessary to go outside a species to obtain viral resistance. *Cucurbita pepo* L. is the most susceptible to all diseases and the most diverse in horticultural types. It includes the "summer squashes" such as Zucchini, eaten while immature, the Acorn and Delicata types, eaten as mature or winter squashes, ornamental and pie pumpkins, and ornamental gourds. *Cucurbita moschata* Poir. is represented mainly by 'Butternut', the most widely grown winter squash in the United States. 'Butternut' is seldom damaged much by CMV but is very susceptible to ZYMV. *Cucurbita maxima* Duch., with its Buttercup and Hubbard types, is probably somewhat similar to *C. moschata* in viral reactions.

We have seen for many years an acute need for resistance to cucumber mosaic in *C. pepo*. Martin (1959) screened many plant introductions (PIs) of that species and found some resistance in PI 176959 from Turkey. Some summer squash with resistance from that source may be close to release by seed companies. Our own efforts to use that resistance continued, but results were disappointing. The best lines showed good resistance in some plantings but almost none in others. We assumed that early infection caused severe stunting, but in hindsight we realize that stunting may have been caused partly by other viruses, of whose existence we were then unaware.

T. W. Whitaker (personal communication, 1972) changed the course of our squash breeding by pointing out that *Cucurbita martinezii* had the highest powdery mildew resistance in the genus. We undertook to transfer this to *C. pepo* by using *C. moschata* as a bridge species. Much of the initial work was done by Max Contin for his Ph.D. thesis (1978). We soon observed that *C. martinezii* and its hybrid with 'Butternut' were almost unaffected by CMV when surrounding summer squash was highly stunted and mottled. Following this work, Provvidenti studied the feral *Cucurbita* species and found resistance to a number of viral diseases (Provvidenti et al. 1978). We have made considerable progress in transferring powdery mildew and cucumber mosaic resistance separately from *C. martinezii* to summer squash (Contin and Munger 1977; Munger and Washek 1983; Washek 1983;), but the next step of combining the two has been severely hampered by natural infection of ZYMV.

Provvidenti found ZYM resistance in both *Cucurbita ecuadorensis* and in *C. moschata*, and we have attempted to transfer it from both species to *C. pepo*. Our tentative conclusion is that ZYM resistance can be transferred rather readily within the species *C. moschata* with probably one main partially dominant gene for resistance (Munger and Provvidenti 1987). There appears to be a reduction in susceptibility when what is presumably the same gene is transferred to *C. pepo*, but we have not yet been able to get an adequate level of resistance to early infection. It is clearly much harder to introduce resistance into a bush variety of squash than into a vine type.

There seems to be general agreement that in squash, even more than in cucumbers, there is need for multiple virus resistance. In both cucumber and the *Cucurbita* species, we frequently find resistance to several viruses together. This fortunate situation is intriguing and hard to explain. It is also in sharp contrast to the situation in *Cucumis melo*, where resistance to one virus is commonly associated with extreme susceptibility to another.

Melons (*Cucumis melo* L.)

Melons belong to a single species within which there are several rather distinct groups, originally called botanical varieties by Naudin. These groups have some characteristic disease reactions and are readily intercrossed. The largest group is the muskmelon/cantaloupe, designated originally as *C. melo reticulatus* and *C. melo cantaloupensis*. The true cantaloupe is grown mostly in Europe, but the name has been widely applied to netted melons in the United States;

the flesh characteristics have always been similar, and the original distinction of warted or scaly versus netted rind has been blurred by hybridization. Today they comprise essentially one group. Next in importance is the *inodorus* group or winter melons, which usually have white or green flesh, smooth or corrugated rind, longer keeping quality, and less musky flavor. Honeydew and Casaba are in this group. It is generally the most susceptible to viral diseases, but resistance to *Fusarium* has come from it.

The *conomon* group includes the oriental pickling melon and some small, sweet, non-netted melons grown in China. This group generally is the most resistant to cucumber mosaic. The *flexuousus* group has many common names such as snake melon, snake cucumber, and Armenian cucumber. It has apparently not been used much if at all in hybridization with other groups in the United States. Some melons from India do not fit any of the groups commonly listed in western publications but have been extremely important sources of resistance to mildew, aphids, and potyviruses.

The first work to my knowledge on mosaic resistance in melons was that of Enzie (1943) at the New York State Agricultural Experiment Station in Geneva, New York. He reported 'Freeman Cucumber', an oriental pickling melon (*C. melo conomon*), as resistant to cucumber mosaic. He crossed it with an unreleased melon of his own breeding named 'Aristogold' and made one backcross to 'Aristogold'. After he left the station, the material was sent to Ithaca in about 1948, where a backcross program continued in which 'Iroquois' and later 'Delicious 51' were used as recurrent parents. Some resistance seemed to be lost with each successive backcross, and therefore several of the better lines were intercrossed to recombine the different genes for mosaic resistance as had been done with cucumber. The outcome with melons was less successful. Four lines judged as best in 1968 did not have a truly high level of mosaic resistance, although in experimental plantings they frequently produced crops where mosaic was severe and commercial varieties failed completely. Monoecious flowering and resistance to races one and two of powdery mildew were backcrossed into these four lines. PMM339 and, to a lesser extent, PMM328 have unusually good shape and depth of flesh for monoecious melons and less autogenic necrosis associated with powdery mildew resistance than most melons. They have shown good combining ability with a number of pollen parents, but unfortunately their hybrids have not shown useful levels of mosaic resistance.

Apparently a much higher level of mosaic resistance is required in melons than in cucumbers. The melon must maintain a healthy vine until fruit matures, whereas cucumbers are harvested immature. Repeated picking of cucumbers tends to delay the senescence that occurs in melons with the ripening of a heavy fruit load. Viral diseases are most devastating on melons when infection occurs as the fruit begins to ripen, and can kill the plant within a few days without the development of obvious symptoms of mosaic. If one is watching the vines closely during this period, subtle symptoms of viral infection can sometimes be seen for a short period before the vines collapse and die. It should be noted, however, that low soil temperatures and other diseases such as powdery mildew can also be associated with this problem.

Continuous Backcrossing for Cucumber Mosaic Resistance

In 1983, to increase cucumber mosaic resistance in the PMM series of melons, we initiated a rapid backcross program patterned after the one used in breeding 'Poinsett 83'. In other words, these melons were crossed again with 'Freeman Cucumber', the original source of CM resistance. Each F_1 was backcrossed to its respective recurrent parent, and mosaic resistance selected in the BC1. We then made a second backcross of the best plants. Third-backcross F_1 progenies were grown in the field in 1985 and had segregates that appeared as resistant as the original 'Freeman Cucumber' F_1 rows. During the pollination period, zucchini yellow mosaic came into the field and eliminated the planting completely before any seed ripened. However, pollen from one good plant had been used to make a fourth backcross onto PMM328 in another field. It is hardly fair to judge the success of this cycle of rapid backcrossing by the one progeny not eliminated by ZYMV, but its advanced generations now designated as CPM328 have a decidedly higher level of mosaic resistance than the PMM328, which we now consider a less resistant check. Two additional backcrosses have been made to PMM339 to improve fruit type.

Anticipating the possible need for a higher level of mosaic resistance than the second backcross cycle, we crossed CPM328 with 'Freeman Cucumber' to begin the third cycle of backcrossing.

Resistance to Potyviruses

For the last backcross in the second cycle, we have used as recurrent parent PPM339, which is PMM339 with the single gene for papaya ringspot resistance transferred by a series of backcrosses from WMR29 (Bohn et al. 1980). More importantly, we are well along with backcrossing a single, incompletely dominant gene for ZYM resistance (Pitrat and Lecoq 1984) into both CPM328 and PPM339. The donor of this resistance was a plant selected by Provvidenti as PI 414723-4. This should soon provide recurrent parents for the CMV improvement program that combine ZYM resistance with PR resistance and fruit type nearly equivalent to PMM339. The 1985 disaster with zucchini yellow mosaic showed that breeding a recurrent parent with ZYM resistance would contribute greatly to progress with CM resistance.

When melon breeders met in November 1987, there seemed to be a consensus that there was no satisfactory source of resistance to WMV, although some resistance had been suggested by the work of Moyer et al. (1985). We therefore initiated a screening of 100 lots of melon germplasm in early 1988. Although none of the germplasm was found to have immunity, some plants with the ability to grow and produce fruit, while checks did neither, were found in four distinct types of melons. These included 'Freeman Cucumber', a wild monoecious melon of the *dudaim* group from Louisiana, PI 371795 from India, and PI 182938, a small, hard melon with no edible flesh from India. We had reason to think that most of these plants might be heterozygous for resistance and that selfing could give higher levels of resistance. That appears to be the case. Likewise, crosses between the selected 'Freeman Cucumber' and selected 371795 gave progenies with higher levels of resistance. A backcross program to add WMV resistance to several varieties has been initiated, but the inheritance of resistance and the degree that can be transferred readily are not yet clear. Progenies with three successive backcrosses to each of four varieties have had numerous segregates with milder symptoms, superior growth, and greater longevity than their recurrent parents when all were inoculated as seedlings with WMV. This experience suggests that inheritance of resistance may be relatively simple and that there is enough dominance to permit reliable selection of plants heterozygous for resistance.

The importance of WM resistance is also unclear. WMV is perhaps the most widespread of the four viruses but the least drastic in its effects. Inoculated plants must grow much longer before resistance to this virus can be evaluated as compared to any of the other three. Commercial plantings with resistance to the other three viruses may be required to reveal the importance of WMV.

The single genes for resistance to papaya ringspot and zucchini yellow mosaic have sufficient dominance that a hybrid with one resistant parent may have satisfactory resistance. However, satisfactory hybrids with CMV resistance will probably require both parents to be resistant and the same may hold for WMV. In order to have pollen parents with CM resistance to make hybrids with a future multiple-virus-resistant version of PMM339, we have undertaken backcross programs in which the recurrent parents are 'UC Topmark' and 'TamUvalde'. To facilitate this CM resistance program, we are developing monoecious and ZYM-resistant versions of the two varieties. Ideally, we should have started these sooner but expect them to overtake the CM-resistant backcrosses, and from then on we will use them as recurrent parents. We do not expect this cycle of backcrossing to yield the level of mosaic resistance we would ultimately like to have, and therefore we expect to initiate another cycle with 'Freeman Cucumber' as a source of higher resistance. It should not take many generations to get a 'Topmark' with monoecious flowering, ZYM resistance, and a modest level of CM resistance. With this as a starting point, a second cycle to raise the level of CM resistance should proceed rapidly and predictably. If monoecism is not desired in the end product, the last backcross in each cycle can be made to the andromonoecious variety to restore the original flowering type.

It is now over 50 years since Enzie discovered a good source of resistance to cucumber mosaic in melons, and yet no material available commercially carries this resistance. There is great urgency to remedy this situation. With the new breeding methodology and germplasm available, we can anticipate a revolution before many years in melon varieties similar to that which has occurred in cucumbers.

Discussion and Conclusions

The high levels of resistance to cucumber mosaic in the 'Tablegreen', 'Marketmore', and 'Poinsett 83' series of cucumbers may be attributable to several factors. (1) The relatively low temperatures of Ithaca, New York favor severity of cucumber mosaic and permit rigorous selecting for resistance. (2) Isolates of CMV that give the most pronounced symptoms have been used, most recently, Zitter's "Fast Strain". In our experience with various CMV isolates, the ranking of lines with different resistance levels is the same, but differences among lines are more pronounced with some isolates. (3) Inoculation of cucumbers is delayed until flowering approaches because plants with little or no resistance are likely to be killed rather than survive with stunted growth as they do when inoculated while young. In *Cucurbita pepo*, however, the opposite situation exists: certain genotypes such as Zucchini are more difficult to infect as they get older.

The alternation of field and greenhouse plantings has not only speeded progress, but each has had its own advantage. The greenhouse generation has permitted evaluation of resistance to a single virus without the interference of other viruses that infect the field plantings sporadically. Field plantings, however, permit a more realistic appraisal of resistance levels and the effects of other pathogens.

The backcross method used so extensively has proven useful in several ways. (1) It has permitted the recovery of recurrent parent characteristics that are impossible to evaluate in segregating material, such as the tropical adaptation of 'Poinsett' and the high proportion of marketable fruit of 'Marketmore' cucumbers. (2) It has speeded breeding by permitting overlapping generations. Recurrent parents can be planted ahead of segregating progenies to have female flowers when the latter produce their first males. (3) As compared with a conventional genetic study as a basis for breeding strategy, backcrossing repeatedly may give even better genetic information while the character is simultaneously being transferred to a useful background. If resistance, for example, can be retained with selection in backcross populations of manageable size, the effect of one or two major genes can be evaluated better as isogenicity approaches than in F_2 and early backcross progenies.

Biotechnology promises to facilitate certain aspects of cucurbit breeding, but when or in what ways remains to be seen. In allocating effort to the newer approaches, it should not be overlooked that there have been recent advances in more conventional plant breeding techniques that make progress more rapid and predictable.

Literature Cited

Barham, W. S., and N. N. Winstead. 1959. 'Ashe' and 'Fletcher', two downy mildew and scab resistant cucumbers. North Carolina Agr. Exp. Sta. Bull. 409.

Barnes, W. C. 1961. Multiple disease resistant cucumbers. Proc. Amer. Soc. Hort. Sci. 77:417–423.

Bohn, G. W., A. N. Kishaba, and J. D. McCreight. 1980. WMR29 muskmelon breeding line. HortScience 15:539–540.

Contin, M. E. 1978. Interspecific transfer of powdery mildew resistance in the genus *Cucurbita*. Ph.D. Thesis, Cornell University.

Contin, M. E., and H. M. Munger. 1977. Inheritance of powdery mildew resistance in interspecific crosses with *Cucurbita martinezii*. HortScience 12:29 (Abstr.).

Enzie, W. D. 1943. A source of muskmelon mosaic resistance found in the oriental pickling melon, *Cucumis melo* var. *conomon*. Proc. Amer. Soc. Hort. Sci. 43: 195–198.

Lane, D. P., and H. M. Munger. 1983. *Ulocladium cucurbitae* leafspot on cucumber (*Cucumis sativus*). Cucurbit Genetics Coop. Report 6:14–15.

Lane, D. P., and H. M. Munger. 1985. Linkage between *Corynespora* leafspot resistance and powdery mildew susceptibility in cucumber (*Cucumis sativus*). HortScience 20:115 (Abstr.).

Martin, M. W. 1959. Inheritance and nature of cucumber mosaic virus resistance in squash. Ph.D. Thesis, Cornell University.

Moyer, J. W., G. G. Kennedy, and L. R. Romanow. 1985. Resistance to watermelon mosaic virus 2 multiplication in *Cucumis melo*. Phytopathology 75:201–205.

Munger, H. M., and R. Provvidenti. 1987. Inheritance of resistance to zucchini yellow mosaic virus in *Cucurbita moschata*. Cucurbit Genetics Coop. Report 10:80–81.

Munger, H. M., and R. L. Washek. 1983. Progress and procedures in breeding CMV resistant *C. pepo* L. Cucurbit Genetics Coop. Report 6:82–83.

Pitrat, M., and H. Lecoq. 1984. Inheritance of zucchini yellow mosaic virus resistance in *Cucumis melo* L. Euphytica 33:57–61.

Porter, R. H. 1929. Reaction of Chinese cucumbers to mosaic. Phytopathology 19:85 (Abstr.).

Provvidenti, R. 1985. Sources of resistance to viruses in two accessions of *Cucumis sativus*. Cucurbit Genetics Coop. Report 8:1.

Provvidenti, R., R. W. Robinson, and H. M. Munger. 1978. Resistance in feral species to six viruses infecting *Cucurbita*. Plant Disease Reporter 62:326–329.

Shifriss, O., C. H. Myers, and C. Chupp. 1942. On resistance to the mosaic virus in cucumber. Phytopathology 32:773–784.

Strong, W. J. 1930. Breeding experiments with the cucumber (*Cucumis sativus* L.). Sci. Agr. 11:333–346.

Walker, J. C., C. F. Pierson, and A. B. Wiles. 1953. Two new scab-resistant cucumber varieties. Phytopathology 43:215–217.

Washek, R. L. 1983. Cucumber mosaic resistance in summer squash (*Cucurbita pepo* L.). Ph.D. Thesis, Cornell University, Ithaca, NY.

Wilson, J. M. 1968. The relationship between scab resistance and fruit length in cucumber, *Cucumis sativus* L. M. S. Thesis, Cornell University, Ithaca, NY.

Zitter, T. A., and L. W. Hsu. 1990. A leaf spot of cucumber caused by *Ulocladium cucurbitae* in New York. Plant Disease 74:824–827.

CHAPTER 3

Breeding Lettuce for Viral Resistance

RICHARD W. ROBINSON & ROSARIO PROVVIDENTI

Introduction

Lettuce (*Lactuca sativa* L.) is a member of the Compositae (Asteraceae) family, tribe Cichorieae (Lactuceae), subtribe Crepidinae. Of the more than 100 species named in the genus *Lactuca*, lettuce has been crossed with only three—*L. serriola*, *L. virosa*, and *L. saligna*. Lettuce and the species compatible with it have nine pairs of chromosomes, but some other *Lactuca* species have *n* of 8 or 17.

Lettuce and *Lactuca* derive their names from a characteristic of the plant, its latex or milky sap (Boswell 1957). The latex was believed in ancient times to be a tranquilizer, and was used for insomnia. A drug known as "lettuce opium" is still produced today from the latex of *Lactuca virosa* (Proula 1985).

Lettuce originated in the Meditteranean region. It has been cultivated since at least 4500 B.C. (Lindqvist 1960), and was served to Iranian rulers of the 6th century B.C. and Romans of the time of Christ (Boswell 1957). By the 5th century A.D. lettuce had become an important vegetable in China and many other countries distant from its center of origin. Lettuce was introduced to the New World by

Drs. Robinson and Provvidenti are in the Departments of Horticultural Science and Plant Pathology, respectively, Cornell University, New York State Agricultural Experiment Station, Geneva, New York 14456.

Columbus in 1494. The original cultivated lettuce was nonheading and quickly formed a seedstalk. Head lettuce was first reported in 1543 (Helm 1954). Today, lettuce is the most popular salad crop throughout the world.

Most of the lettuce production in the U.S. throughout the year is in California; Arizona, Florida, and Texas are also important production areas in the winter. Growers in New York, New Jersey, and midwestern states market lettuce in summer months. Crisphead (iceberg) varieties predominate in U.S. markets, especially for long-distance shipments, but romaine (cos), butterhead, and leaf types are also sold. Some Bibb-type lettuce is being grown in greenhouses now, often hydroponically, but the majority of lettuce in U.S. supermarkets is field grown.

Until this century, lettuce breeding was simply by selection within existing varieties. The use of controlled pollinations by Jagger et al. (1941) and others led to significant improvements, including disease resistant varieties. Lettuce varieties resistant to fungal diseases were introduced nearly half a century before the first virus resistant lettuce variety (Ryder 1979b). Recently, there has been increasing interest in breeding lettuce for virus resistance.

The primary objective of the lettuce breeding program at Cornell University is multiple virus resistance. Viruses are a major problem for lettuce growers not only in New York but in most other areas of the world. In some years hundreds of acres in New York are not harvested because high incidence of virus infection makes the crop unmarketable. The problem in New York is especially acute with the fall crop due to high aphid populations. In the spring crop, viral infection is often sporadic, but it increases during the season and reaches a peak in the fall.

Virus Resistance in Lettuce

The following definitions are used in the discussion of viral diseases of lettuce. Plants that remained symptomless and upon assays were determined to be locally and systemically free of viral infection were rated as *highly resistant* or *immune.* Plants in which viral infection was confined to the inoculated leaves were classified as *resistant.*

Plants that developed mild and transitory systemic symptoms or were symptomless carriers of the virus were considered to be *tolerant* or *highly tolerant.* Plants exhibiting persistent moderate to severe symptoms were regarded as *susceptible.*

Lettuce Mosaic Virus

Lettuce mosaic virus (LMV) is a virus with flexuous rod particles about 750 × 13 nm, containing a single strand of infectious RNA. It is spread in nature by several species of aphids in the nonpersistent manner. Because it can be seed transmitted in lettuce, this virus is of world-wide distribution. LMV can infect species of more than 20 genera in 10 families of plants (Tomlinson 1970b).

LMV was recognized in the 1950s as a serious disease of lettuce in the Salinas Valley of California (Ryder 1968), and was the first virus disease to engage the interest of lettuce breeders. Ryder (1968) screened nearly a hundred different varieties and many breeding lines of lettuce, but all were susceptible. He used radiation and ethyl methane sulfonate as mutagenic agents, but was unable to induce resistance to lettuce mosaic. However, when he tested 142 USDA plant introductions of *Lactuca* he found three that were resistant. Each had originated in Egypt and had been classified as *Lactuca serriola*, a very close relative of lettuce. Ryder, however, concluded they were primitive forms of *Lactuca sativa*, the same species as cultivated lettuce.

Von der Pahlen and Crnko (1965) had previously found that 'Gallega', a romaine-type lettuce variety from Spain, is resistant to LMV. Ryder (1970a) discovered that the variety 'Fordhook' is resistant to LMV, and concluded that 'Fordhook' is synonymous to 'Gallega'. Bannerot et al. (1969) determined that the resistance of 'Gallega' is due to a single recessive gene. Ryder (1970b) found that 'Fordhook' and an Egyptian introduction also have a single recessive gene for resistance, which is the same as or is allelic to the one in 'Gallega' and two other resistant Spanish varieties. Bannerot et al. (1969) used the symbol g for this gene, but the symbol *mo* proposed by Ryder (1970b) is preferred (Robinson et al. 1983) due to the prior use of the symbol g for another gene.

The *mo* gene appears to reduce drastically the rate of viral replication and restrict the systemic movement of the virus in the lettuce plant (Ryder and Whitaker 1980). However, inoculated resistant plants may develop systemic infection and the virus can be transmitted through their seed, although at a much lower rate than in susceptible plants (Ryder and Whitaker 1980, Ryder 1973, 1964). This possible seed transmission had led some researchers to be reluctant to use the *mo* gene in lettuce breeding, in fear that a resistant variety might serve as a carrier and transmit the virus to susceptible varieties. However, no problems have been reported in the 15 years that LMV resistant varieties have been available in the US. Although resistant plants may harbor the virus without displaying prominent symptoms, LMV can be detected by ELISA or indicator host tests.

Lettuce mosaic is the only important virus disease of lettuce that is seed transmitted. It can also be transmitted through pollen, but at a lower frequency than ovule transmission (Ryder 1973, 1964). LMV is often introduced into a lettuce field by a few infected seeds, and then is spread by aphids. Grogan et al. (1952) developed a very useful method of reducing losses from LMV by testing seedlots to determine the incidence of seedborne infection. By the use of seed certified to be virtually free of LMV (0 in 30,000 seeds tested), growers have been able to reduce greatly their losses from this virus. Our surveys taken in the lettuce fields of New York for several years indicate a steady decline in the importance of LMV corresponding with the increase in use of indexed seed.

The use of virus-free seed has greatly reduced but not completely eliminated the problem caused by lettuce mosaic. The disease still occurs in New York and elsewhere, hence there is merit in breeding resistant varieties to complement the seed indexing program. Accordingly, LMV resistance is one of the objectives of the lettuce breeding program at Geneva, New York. We have used 'Vanguard 75', developed by backcrossing *mo* into 'Vanguard' (Ryder 1975), as our source of LMV resistance.

The resistance to LMV conferred by the *mo* gene is strain-specific. Zink et al. (1973) found a strain (LMV-L) that is virulent on 'Gallega' and other varieties with this gene. The response of different lettuce varieties to infection with the LMV-L strain varied. Some varieties developed nonlethal systemic mosaic, but others developed a severe necrosis of the leaf blades and many of these plants died. Interestingly, all of the varieties with the lethal reaction are resistant to downy mildew and susceptible to turnip mosaic virus. In crosses between Great Lakes 118 (lethal reaction) and 'Calmar' and 'Imperial 410' (nonlethal), F_2 ratios of 9 nonlethal to 7 with the lethal reaction to LMV-L indicated segregation for two dominant complementary genes.

Fortunately, no evidence could be found that this virulent strain of LMV is seed-transmitted, in contrast to the original strain of LMV. This apparent lack of seed transmission may be responsible for the relative unimportance of the virulent LMV strain in most commercial lettuce fields. Although it was found 20 years ago in a severely infected field in Salinas, California, where it was associated with the host Bristly Oxtongue (*Picris echiodes* L.) growing as a weed there, it has not been reported as occurring frequently elsewhere since then. More recently, there was concern when an unusually severe strain of LMV occurred in the Salinas Valley in 1986 and 1987 (Pryor 1987), but breeding lines with the *mo* gene proved to be resistant to this severe strain as well as to the common strain of lettuce mosaic virus. Thus, the continued use of the *mo* gene by lettuce breeders seems justified.

Lettuce varieties differ in their rate of seed transmission of LMV. Kassanis (1947) reported that the susceptible variety 'Cheshunt Early Giant' does not transmit the virus through its seed when infected. Genetic variation apparently occurs within the variety 'Bibb' for rate of seed transmission (Couch 1955). J. E. Welch selected in his lettuce breeding program at the University of California, Davis for plants that would not transmit LMV through their seed, although susceptible to the virus, as another form of genetic control of lettuce mosaic virus (Ryder 1979b). The inheritance of genetic difference in the rate of seed transmission of LMV has not been reported.

Turnip Mosaic Virus

Turnip mosaic virus (TuMV) has typical filamentous rod shaped particles approximately 720 nm long containing a single strand of RNA (Tomlinson 1970a). It is mechanically transmitted experimentally, but in nature is efficiently spread by numerous aphid species in the nonpersistent manner. It can be found wherever wild and cultivated crucifers are growing, but it can also infect many species of other genera, including *Lactuca*. Apparently this virus is not transmitted via seeds.

Turnip mosaic virus is seldom a problem in lettuce, since most varieties are resistant. It was not until the variety 'Calmar' was developed, by transferring from *L. serriola* a single dominant gene for resistance to downy mildew, that growers found their fields of this variety infected with a disease they had never seen before. Research by Zink and Duffus (1969) revealed that the single dominant gene for susceptibility to turnip mosaic virus is associated with resistance to downy mildew. When the breeder transferred the gene from *L. serriola* for resistance to race 5 of *Bremia lactucae*, he inadvertently introduced a gene for susceptibility to TuMV, which is very closely linked to the gene for downy mildew resistance. Zink and Duffus found that not all accessions of *L. serriola* are susceptible to TuMV, but when susceptibility to TuMV occurs in *L. serriola* or *L. sativa*, it is associated with resistance to downy mildew.

Bidens Mottle Virus

Bidens mottle virus (BiMV) causes an important disease of lettuce and endive in Florida. Like other potyviruses, BiMV has flexuous virus rods about 720 nm long, containing a single strand of RNA (Purcifull et al. 1976). It is transmitted by a number of aphid species in the nonpersistent manner, and also by mechanical means. There is no indication that this virus is seed transmitted.

Zitter and Guzman (1974) found that the variety 'Valmaine' is resistant to bidens mottle virus, due to a single recessive gene for which they (1977) gave the symbol *bi*. Strain specificity occurs for resistance to BiMV; an isolate of this virus was found to overcome the resistance conferred by *bi* (Zitter and Guzman 1977).

Broad Bean Wilt Virus

Broad bean wilt virus (BBWV) also occurs in New York lettuce fields (Atilano 1971; Bruckhart and Lorbeer 1975, 1976; Provvidenti et al. 1984). BBWV also occurs on lettuce in Europe, Japan, and Australia (Taylor and Stubbs 1972). The isometric particles of this virus are about 25 nm in diameter and contain a single strand of RNA (Taylor and Stubbs 1972). It is spread by several aphid species in the nonpersistent manner, but it is also sap-transmitted. The virus can infect a very large number of plant species, including several cultivated crops of major economic importance. Efforts to demonstrate seed transmission of BBWV have been unsuccessful.

All of the major lettuce varieties grown in New York when our lettuce breeding program began in 1975 are susceptible to broad bean wilt virus. We undertook a search for resistance, and found that several but not all introductions of *Lactuca virosa* are resistant. This species can be crossed with lettuce (Thompson et al. 1941), but the hybrid is highly sterile. Before undertaking such a difficult breeding program with *L. virosa*, we tried to locate resistance in *L. sativa*, and found that 28 of the 46 lettuce varieties tested were resistant to BBWV (Provvidenti et al. 1984). Paradoxically, most of the lettuce varieties grown in California (where BBWV is seldom serious) are resistant to that virus, and many European varieties are also resistant, yet most of the important lettuce varieties grown then in New York (where BBWV often occurs) are susceptible. The only resistant crisphead variety grown in New York at the time of our survey was 'Stokes Evergreen'.

There seems to be no logical explanation for this situation, of growing susceptible varieties in an area where a disease is serious but resistant varieties where the disease is unimportant. Coincidence and the rather narrow genetic base of most lettuce grown in the Northeast at that time are probably responsible. 'Minetto', 'Fulton', and 'Oswego', popular lettuce varieties in New York then, are all derived from crosses between 'Empire' and 'Cornell 456' (Raleigh 1964). The susceptible variety 'Fulton' is a parent of other susceptible varieties grown in New York, including 'Fairton' and 'Ithaca'. All of the crisphead varieties bred at Cornell University, which were then the leading varieties grown in the Northeast, Midwest, and Florida, are susceptible to BBWV. 'Stokes Evergreen', the only BBWV resistant variety grown commercially to any extent in the Northeast then, has a different origin. It is a selection from the resistant variety 'Great Lakes' (J. F. Gale 1983, Stokes Seed Co., personal communication).

It was fortunate for our breeding program that 'Vanguard 75' is among the many western varieties that are resistant to BBWV. We originally used this variety as a source of resistance to LMV, but later discovered that 'Vanguard 75' is resistant to BBWV as well (Provvidenti et al. 1984). The original 'Vanguard' variety is resistant to

BBWV. When Ryder (1975) backcrossed the gene for LMV resistance into that variety, he fortuitously combined resistance to two viral diseases of lettuce for the very first time.

Not until several years after we used 'Vanguard 75' as a parent for its LMV resistance did we find that it is also resistant to BBWV. By that time, we had developed numerous pedigree selection and backcross lines from crosses between 'Vanguard 75' and various susceptible varieties bred at Cornell University. Previously, the breeding lines had been selected only for LMV resistance, but when we tested them for BBWV some were found to be resistant to both viruses. Thus, several years were saved in the breeding program by inadvertently using a parent resistant to both BBWV and LMV.

Our tests indicate that a single gene for BBWV resistance distinguishes 'Vanguard 75' from susceptible varieties. No linkage was detected between this gene for BBWV resistance and the genes of 'Vanguard 75' for LMV resistance, scalloped leaf margin (Ryder 1965), or resistance to downy mildew (Yuen and Lorbeer 1983). In crosses among 'Calmar', 'Vanguard 75', 'Stokes Evergreen', and other resistant varieties, all appeared to have the same gene for resistance to BBWV, since only resistant plants were found in the F_2.

Lettuce Big Vein Virus

Lettuce big vein virus (LBVV) is of much importance in California, and also occurs in other areas. It is a very difficult disease to work with, but tolerant varieties have been identified (Campbell et al. 1964; Ryder 1979a). Ryder (1986) has released LBVV-resistant breeding lines.

Recent information (Vetten et al. 1987) has indicated that LBVV has very labile rod-shaped particles varying in length from 324 to 152 nm with a width of 18 nm. In nature it is transmitted by the soil-borne fungus *Olpidium brassicae* (Campbell et al. 1963). It is also transmissible by grafting but not by sap. Serologically it is related to tobacco stunt virus.

Beet Western Yellows Virus

Lettuce varieties also differ for susceptibility to beet western yellows virus (BWYV). Butterhead-type varieties are especially susceptible (Watts 1975), but apparently no variety is highly resistant. Walkey and Bolland (1987) and Walkey and Pink (1990) found varietal differences in tolerance; crisphead and 'Batavian' types were most tolerant. The highest degree of resistance they found was in wild *Lactuca* species that cannot be crossed with lettuce. Ashby et al. (1979) found that the lettuce varieties 'Meikoningen' and 'Reskia' had a low incidence of BWYV in the field, but they considered that it might be due to aphid avoidance or delayed symptom expression.

This luteovirus possesses icosahedral particles about 26 nm in diameter (Duffus 1977). Under natural conditions, it is spread by aphid species in a persistent (circulative) manner but is not transmissible by mechanical means. The virus can be found in many countries of the world and has a very wide host range.

Lettuce Necrotic Yellows Virus

Lettuce necrotic yellow virus (LNYV) is widespread in Australia and New Zealand, and has also been reported in other countries. Nguyen (1986) has identified several lettuce varieties, including 'Bibb', as promising sources of resistance. LNYV is a rhabdovirus with bacilliform particles about 227 × 66 nm containing a single strand of RNA. In nature it is spread by aphid species in a persistent manner; experimentally it can be transmitted mechanically (Francki and Randles 1970).

Lettuce Mottle Virus

Lettuce mottle virus (LMoV) is a recently described virus occurring in Brazil, where it seems to be widespread. The virus particles are isometric, measuring about 30 nm in diameter. It causes mottling or mosaic in many lettuce varieties but apparently is not seed transmitted. Under field conditions it is spread in the nonpersistent manner by the aphid *Hyperomzus lactucae.* Experimentally, five varieties were found to be resistant to this virus: 'Guape', 'Monte Allegre', 'Piracicaha', 'Prize Head', and 'Romana Branca' (Marinho and Kitajima (1986).

Cucumber Mosaic Virus

Cucumber mosaic virus (CMV) is the type member of the cucumovirus group. The icosahedral particles contain 3 functional pieces of a single-stranded RNA (Francki et al. 1979). It is spread by more than 60 aphid species in the nonpersistent manner, and can also be transmitted mechanically. CMV has a very wide host range, more than 800 plant species. It can be seed-transmitted in at least 19 plant species, but not in lettuce. Although CMV is of world-wide distribution, it seems to be more prevalent in the temperate regions. In New York, CMV is the most serious virus disease of lettuce, followed by BBWV and LMV, in that order of importance (Provvidenti et al. 1984). The incidence of CMV, as with other diseases, varies from year to year. In the late 1970s, it was not uncommon to see many commercial fields of lettuce in New York abandoned because of 100% infection with CMV. In recent years, CMV has been less prevalent but still remains an important factor in lettuce production.

No source of CMV resistance was known when the lettuce breeding program at Geneva, New York was initiated in 1975. Accordingly, we searched for resistant germplasm by testing many varieties and breeding lines of lettuce and 566 USDA Plant Introductions of *Lactuca* from all over the world. Without exception, every accession of *Lactuca sativa* and *L. serriola* tested was susceptible to CMV. However, a more distantly related species, *L. saligna*, was found to be resistant (Provvidenti et al. 1980). Several of the plant introductions that had been classified by the USDA as being *L. saligna* were susceptible, but they had been misclassified and actually were not *L. saligna*. Only *L. saligna* PI 261653 from Portugal was resistant to CMV.

The interspecific cross between *Lactuca saligna* and *L. sativa* was first reported by Thompson et al. (1941). According to the literature, unilateral incompatibility exists and the cross can be made only with lettuce as the female parent, but we also succeeded in making the reciprocal cross (R. W. Robinson and R. Provvidenti, unpublished). The interspecific hybrid has a high degree of sterility, but by growing many interspecific hybrid plants we were able to produce sufficient seed to make possible the very large F_2 populations needed. There was a preponderance of undesirable types and sterility in early segregating generations, but by rigid selection for desirable horticultural type and fertility as well as for CMV resistance, we were able to develop the variety 'Saladcrisp'. 'Saladcrisp', derived by pedigree selection from 'Ithaca' × *L. saligna* PI 261653, is the only lettuce variety which is resistant to CMV.

When NY 18-78, the breeding line that was later named 'Saladcrisp', was included in variety trials in growers' fields for several years, it developed symptoms of mosaic in one of those trials. Since LMV, BBWV, and CMV can all produce similar symptoms on lettuce, and the breeding line was known to be resistant to only CMV, at first it was presumed that the disease in the field was caused by one of the other viruses. Tests with indicator plants, however, established that the virus in the field was a unique pathotype of CMV. Tests with *L. saligna* PI 261653 confirmed that its CMV resistance is strain-specific. Edwards et al. (1983) found differences in RNA composition of CMV pathotypes differing for virulence to 'Saladcrisp'.

Until a source of CMV resistance in *Lactuca* was found, there was no way to determine that different strains of the virus were present. Our experience emphasizes the importance of close cooperation between a virologist and a breeder in breeding for virus resistance. Without the cooperation of the virologist, the breeder and lettuce growers might have been surprised if a CMV resistant variety succumbed to that disease soon after its release. By knowing that 'Saladcrisp' is resistant to one but susceptible to another strain of CMV, we were able to alert New York growers to the limitation of this resistant variety when it was released. The occurrence of CMV strains that are virulent to 'Saladcrisp' does not completely negate the value of the resistance of this variety, since the CMV L-1 pathotype to which 'Saladcrisp' is resistant is of widespread occurrence.

The discovery of CMV strains overcoming the resistance of 'Saladcrisp', several years before it was released, saved us time in breeding for resistance to other strains of CMV. We have found tolerance to CMV strain L-2, as well as resistance to the original L-1 strain, in different accessions of *L. saligna*. All accessions of *L. saligna* tested are resistant to pathotype L-1, but some accessions are also tolerant to L-2.

Still another strain of CMV, virulent to those sources of resistance, was found in a commercial field of lettuce near Oswego, New York. All accessions of *L. saligna* tested were susceptible to the L-3 strain of CMV. One accession obtained as *L. saligna* turned out to be a seed mixture containing *L. serriola* as well as *L. saligna*. This mistake was beneficial for our breeding program, for some of the *L. serriola* contaminants proved to be tolerant to L-3 as well as to the L-1 and L-2 strains of CMV.

Unlike L-1 resistance from PI 261653, in which inoculated plants are symptomless and have no systemic spread of the virus, L-2 or L-3 tolerant plants do develop a mild systemic mottle when inoculated with those CMV strains. However, the mottling is less intense, and plants do not develop the extreme stunting that characterizes susceptible plants infected with CMV. The known sources of tolerance to the L-2 and L-3 strains appear to be useful for breeding programs.

The differential reaction of sources of resistance in *Lactuca* to different strains of CMV contrasts sharply with the stable CMV resistance of cucurbits. CMV resistant cucumber varieties, deriving their resistance from germplasm developed by H. M. Munger (1993), have been grown for several decades without any report of resistance breaking down due to a new strain of the virus. CMV resistant cucumber varieties, and also squash breeding lines deriving CMV resistance from *Cucurbita martinezii* (Munger 1976), were resistant in our tests with the L-1, L-2 and L-3 strains of CMV.

CMV resistance in cucumber (Shifriss et al. 1942) and squash (Washek 1983) is of complex inheritance, but our tests indicate simple Mendelian inheritance for the CMV resistance of PI 261653. The greater stability of CMV resistance in cucurbits, compared with lettuce, is probably due to the greater number of resistance genes in cucurbits. Quite possibly, greater stability of CMV resistance in lettuce might be achieved by pyramiding different resistance genes.

Despite the occurrence of virus strains that can overcome genes of *Lactuca* for resistance to CMV, LMV, BiMV, and probably other viruses as well, virus resistance in lettuce is likely to be considerably more durable than resistance to fungal diseases such as downy mildew, where numerous races occur and are often found on resistant varieties soon after their release. Since gene recombination following hybridization does not occur in viruses as it does in *Bremia*, there is less likelihood for new virulent strains of viruses overcoming resistant genotypes.

Multiple Resistance of *Lactuca saligna*

Lactuca saligna, which we have used as a source of CMV resistance, has considerably more to offer the lettuce breeder. Resistance to lettuce infectious yellows virus (LIYV) is unknown in *L. sativa* (Dorst et al. 1983), but McCreight et al. (1986, 1987) discovered that PI 261653 and some, but not all other accessions of *L. saligna* are resistant. Lettuce varieties differ in reaction to LIYV, with 'Climax' being the most tolerant of the varieties tested, but none was as resistant as *L. saligna*. LIYV virus particles are very long, rod shaped particles about 1200 to 2000 nm long and 10–12 nm wide, belonging to the closterovirus group (Duffus et al. 1986). Serologically LIYV is not closely related to the other members of the group, such as beet yellows virus. It is of common occurrence in the southwest desert regions of the USA, particularly in the Imperial Valley (California) and Arizona. It is very destructive and is spread by the whitefly, *Bemisia tabaci*, in a semipersistent manner. Individual insects can acquire the virus in a few minutes and retain it for three days.

Tomato spotted wilt virus (TSWV) can be a severe problem for lettuce growers in Hawaii and other areas, but R. Provvidenti found that some accessions of *L. saligna* are resistant to the virus (Cho et al. 1989). Hartmann and O'Malley (1986) reported that some lettuce varieties are tolerant to TSWV, and this tolerance is partially dominant (O'Malley and Hartmann 1989). TSWV is a unique plant virus, since its particles (70–90 nm) are covered by a lipoprotein envelope (Ie 1970). It is also the only lettuce virus transmitted in a circulative manner by certain species of thrips, and is highly unstable *in vitro*. TSWV can infect a very large number of plant species and it is of common occurrence in temperate and subtropical areas of the world.

The first disease resistance reported in *L. saligna* was for downy mildew (Netzer et al. 1976). *L. saligna* has also been reported to be resistant to *Stemphylium* (Netzer et al. 1985) and anthracnose (Ochoa et al. 1987). Conrad (1982) determined that *L. saligna* PI 261653 was resistant to one isolate of the root knot nematode, *Meloidogyne hapla*, but susceptible to another isolate of the same species. Kishaba et al. (1973) reported that *L. saligna* is also resistant to the cabbage looper.

Thus, *L. saligna* is resistant to at least three viruses (CMV, TSWV, and LIYV), a nematode, several fungal diseases (downy mildew, *Stemphylium*, and anthracnose), the bacterial disease of corky root rot, and the cabbage looper insect. It is truly a treasure trove for the lettuce breeder. Yet, despite this plethora of disease and insect resistance, *L. saligna* has been used but very little in lettuce breeding to date. It was nearly half a century after this wild species was first crossed with lettuce that 'Saladcrisp', the first lettuce variety derived from a cross with *L. saligna*, was released.

Future Prospects

'Saladcrisp' is the only lettuce variety with *L. saligna* in its pedigree, and there is still only one variety resulting from the direct cross between *Lactuca virosa* and lettuce. In one of the greatest lettuce breeding accomplishments of all times, R. C. Thompson developed 'Vanguard' from the amphidiploid hybrid of lettuce × *L. virosa* (Thompson and Ryder 1961). Hybridization with distantly related species has been used but very little in lettuce breeding in the past, but is likely to become more important in the future.

Biotechnology has also seldom been utilized by lettuce breeders; all lettuce varieties so far developed have resulted from conventional breeding procedures. An attempt to utilize somaclonal variation from tissue culture to breed lettuce for CMV resistance (R. Alconero and R. Provvidenti personal communication) was unsuccessful. Nevertheless, greater use of biotechnology is expected in the future to breed lettuce for virus resistance and other attributes.

Conclusion and Summary

Genetic sources of resistance or tolerance in *Lactuca* have been found for lettuce mosaic virus, turnip mosaic virus, bidens mottle virus, broad bean wilt virus, beet western yellows virus, cucumber mosaic virus, big vein, spotted wilt, lettuce mottle virus, lettuce necrotic yellows virus, and infectious yellows virus. Resistance in several cases is strain-specific, but virus resistance in lettuce has been relatively stable and durable.

Conventional breeding techniques, rather than biotechnology procedures, have been used to date in lettuce improvement.

Interspecific hybridization has been used to incorporate cucumber mosaic virus resistance into lettuce. Susceptibility to turnip mosaic virus was inadvertently transferred to lettuce from *L. serriola* with a closely linked gene for downy mildew resistance.

Conventional breeding, augmented by interspecific hybridization, tissue culture, protoplast fusion, transformation, and other novel techniques, can be expected to develop multiple resistant lettuce varieties in the future. Lettuce growers can look forward to controlling many virus diseases of lettuce by the increased use of resistant varieties.

Literature Cited

Ashby, J. W., L. Bos, and N. Huijberts. 1979. Yellows of lettuce and some other vegetable crops in the Netherlands caused by beet western yellows virus. Neth. J. Pl. Path. 85:99–111.

Atilano, R. A. 1971. Identification of three viruses from New York lettuce growing in organic soil. MS Thesis, Cornell Univ., Ithaca, NY.

Bannerot, H., L. Boulidard, J. Manou, and M. Duteil. 1969. Étude de l'héredite de la tolérance au virus de la mosaique de la laitue chez la variété Gallega de inviernu. Ann. Phytopathology 1:219–226.

Boswell, V. R. 1957. Our vegetable travelers. In The World in Your Garden, Eds. W. H. Camp, V. R. Boswell, J. R. Magness. Washington, D.C.: Natl. Geogr. Soc. 148–149.

Bruckhart, W. L., and J. W. Lorbeer. 1975. Recent occurrences of cucumber mosaic, lettuce mosaic and broad bean wilt viruses in lettuce and celery fields in New York. Plant Dis. Reptr. 59:203–206.

Bruckhart, W. L., and J. W. Lorbeer. 1976. Cucumber mosaic virus in weed hosts near commercial fields of lettuce and celery. Phytopathology 66:253–259.

Campbell, R. N., and R. G. Grogan. 1963. Big vein of lettuce and its transmission by *Olpidium brassicae*. Phytopathology 53:252–259.

Campbell, R. N., R. G. Grogan, and K. A. Kimble. 1964. Big vein of lettuce. Calif. Ag. 18(3):6–8.

Cho, J. J., R. F. L. Man, T. L. German, R. W. Hartmann, L. S. Yudin, D. Gonsalves, and R. Provvidenti. 1989. A multidisciplinary approach to management of spotted wilt virus in Hawaii. Plant Disease 73:375–382.

Conrad, B. 1982. Influence of plant resistance and nematicides on growth and yield of tomato (*Lycopersicon esculentum*) and on population dynamics of root-knot nematodes (*Meloidogyne incognita* and *Meloidogyne hapla*), with an associated study of root-knot nematode resistance in lettuce (*Lactuca* species). Cornell Univ. PhD thesis, Ithaca, NY.

Couch, H. B. 1955. Studies on seed transmission of lettuce mosaic virus. Phytopathology 45:63–70.

Dorst, H. J. M. van, N. Huijberts, and L. Bos. 1983. Yellows of glasshouse vegetables, transmitted by *Trialeuroda vaporariorum*. Neth. J. Pl. Path. 89:171–184.

Duffus, J. E. 1977. Beet western yellows virus. Descriptions of Plant Viruses No. 89. Kew, Surrey, England: Commonw. Mycol. Inst./Assn. Appl. Biol.

Duffus, J. E., R. C. Larsen, and H. Y. Liu. 1986. Lettuce infectious yellows virus: a new type of whitefly-transmitted virus. Phytopathology 76:97–100.

Edwards, M. C., D. Gonsalves, and R. Provvidenti. 1983. Genetic analysis of cucumber mosaic virus in relation to host resistance: location of determinants for pathogenicity to certain legumes and *Lactuca saligna*. Phytopathology 73:269–273.

Francki, R. I. B., D. W. Mossop, and T. Hatta. 1979. Cucumber mosaic virus. Descriptions of Plant Viruses No. 213. Kew, Surrey, England: Commonw. Mycol. Inst./Assn. Appl. Biol.

Francki, R. I. B., and J. W. Randles. 1970. Lettuce necrotic yellows virus. Descriptions of Plant Viruses, No. 26. Kew, Surrey, England: Commonw. Mycol. Inst./Assn. Appl. Biol.

Grogan, R. G., J. E. Welch, and R. Bardin. 1952. Common lettuce mosaic and its control by the use of mosaic-free seed. Phytopathology 42:573–578.

Hartmann, R. W., and P. J. O'Malley. 1986. Tomato spotted wilt virus resistance in lettuce (*Lactuca sativa* L.). HortScience 20:790 (Abstract).

Helm, J. 1954. *Lactuca sativa* in morphologisch-systematischer Sicht. Kulturpflanze 2:72–129.

Ie, T. S. 1970. Tomato spotted wilt virus. Descriptions of Plant Viruses No. 39. Kew, Surrey, England: Commonw. Mycol. Inst./Assn. Appl. Biol.

Jagger, I. C., T. W. Whitaker, J. J. Uselman, and M. M. Owen. 1941. The Imperial strains of lettuce. U.S.D.A. Circ. 596.

Kassanis, B. 1947. Studies on dandelion yellow mosaic and other virus diseases of lettuce. Ann. Appl. Biol. 34:412–421.

Kishaba, A. N. 1973. Differential oviposition of cabbage loopers on lettuce. J. Amer. Soc. Hort. Sci. 98:367–370.

Lindqvist, K. 1960. On the origin of cultivated lettuce. Hereditas 46:319–350.

Marinho, V. L. and E. W. Kitajima. 1986. Lettuce mottle virus, an isometric virus transmitted by aphids. Fitopatologia Brasileira 11:923–936.

McCreight, J. D. 1987. Resistance in wild lettuce to lettuce infectious yellows virus. HortScience 22:640–642.

McCreight, J. D., A. N. Kishaba, and K. S. Mayberry. 1986. Lettuce infectious yellows tolerance in lettuce. J. Amer. Soc. Hort. Sci. 111:788–792.

Munger, H. M. 1976. *Cucurbita martinezii* as a source of disease resistance. Veg. Improv. Newsletter 18:4.

Munger, H. M. 1993. Breeding for viral disease resistance in cucurbits. In Resistance to Viral Diseases of Vegetables: Genetics and Breeding. Ed. M. M. Kyle. Portland, OR: Timber Press.

Netzer, D., D. Globerson, and J. Sacks. 1976. *Lactuca saligna* L., a new source of resistance to downy mildew (*Bremia lactucae* Reg.). HortScience 11:612–613.

Netzer, D., D. Globerson, C. Weintal, and R. Elyassi. 1985. Sources and inheritance of resistance to *Stemphylium* leaf spot of lettuce. Euphytica 34:393–396.

Nguyen, V. Q. 1986. Evaluations for necrotic yellows resistance in lettuce. HortScience 21:789 (Abstr.).

Ochoa, O., B. Velp, and R. W. Michelmore. 1987. Resistance in lettuce to *Microdochium pannatoniana* (lettuce anthracnose). Euphytica 36:609–614.

O'Malley, P. J., and R. W. Hartmann. 1989. Resistance to tomato spotted wilt virus in lettuce. HortScience 24:360–362.

Proula, E. A. 1985. The Fine Art of Salad Gardening. Rodale Press, Emmaus, PA: Rodale Press.

Provvidenti, R., R. W. Robinson, and J. W. Shail. 1984. Incidence of broad bean wilt virus in lettuce in New York State and sources of resistance. HortScience 19:569–570.

Provvidenti, R., R. W. Robinson, and J. W. Shail. 1980. A source of resistance to a strain of cucumber mosaic virus in *Lactuca saligna* L. HortScience 15:528–529.

Pryor, A. 1987. Mosaic miseries. Calif. Farmer 267(6):32–33.

Purcifull, D. E., S. R. Christie, and T. A. Zitter. 1976. Bidens mottle virus. Descriptions of Plant Viruses, No. 161. Kew, Surrey, England: Commonw. Mycol. Inst./Assn. Appl. Biol.

Raleigh, G. J. 1964. 'Oswego', 'Fulton' and 'Minetto' lettuce varieties and their adaptation. Ithaca, NY: Cornell Univ. Veg. Crop. Dept. Mimeo VC122.

Robinson, R. W., J. D. McCreight and E. J. Ryder. 1983. The genes of lettuce and closely related species. Pl. Breed. Rev. 1:267–293.

Ryder, E. J. 1964. Transmission of common lettuce mosaic through the gametes of the lettuce plant. Plant. Dis. Reptr. 48:522–523.

Ryder, E. J. 1965. The inheritance of five leaf characters in lettuce (*Lactuca sativa* L.). Proc. Amer. Soc. Hort. Sci. 86:457–461.

Ryder, E. J. 1968. Evaluation of lettuce varieties and breeding lines for resistance to common lettuce mosaic. U.S.D.A. Tech. Bull. 1391, 8 pp.

Ryder, E. J. 1970a. Screening for resistance to lettuce mosaic. HortScience 5:47–48.

Ryder, E. J. 1970b. Inheritance of resistance to lettuce mosaic. J. Amer. Soc. Hort. Sci. 95:378–379.

Ryder, E. J. 1973. Seed transmission of lettuce mosaic virus in mosaic resistant lettuce. J. Amer. Soc. Hort. Sci. 98:610–614.

Ryder, E. J. 1975. 'Vanguard 75' lettuce. HortScience 14:283–284.

Ryder, E. J. 1979a. Effects of big vein and temperature on disease incidence and percentage of plants harvested of crisphead lettuce. J. Amer. Soc. Hort. Sci. 104:665–668.

Ryder, E. J. 1979b. Leafy Salad Vegetables. In: Breeding Vegetable Crops. Ed. M. J. Bassett. Westport, CT: AVI Publ. Co.

Ryder, E. J., and T. W. Whitaker. 1980. The lettuce industry in California: a quarter century of change, 1954–1979. Hort. Rev. 2:164–207.

Ryder, E. J. 1986. Release of six big vein resistant breeding lines. U.S.D.A. mimeo.

Shifriss, O., C. H. Myers, and C. Chupp. 1942. Resistance to mosaic virus in the cucumber. Phytopathology 32:777–784.

Taylor, R. H., and L. L. Stubbs. 1972. Broad bean wilt virus. Descriptions of Plant Viruses No. 81. Kew, Surrey, England: Commonw. Mycol. Inst./Assn. Appl. Biol.

Thompson, R. C., and E. J. Ryder. 1961. Descriptions and pedigrees of nine varieties of lettuce. USDA Tech. Bull. 1244. 19 pp.

Thompson, R. C., T. W. Whitaker, and W. F. Kosar. 1941. Interspecific genetic relationships in *Lactuca*. J. Agric. Res. 63:91–107.

Tomlinson, J. A. 1970a. Turnip mosaic virus. Description of Plant Viruses No. 8. Kew, Surrey, England: Commonw. Mycol. Inst./Assn. Appl. Biol.

Tomlinson, J. A. 1970b. Lettuce mosaic virus. Description of Plant Viruses No. 9. Kew, Surrey, England: Commonw. Mycol. Inst./Assn. Appl. Biol.

Vetten, H. J., D. E., Lesemann, and J. Dalchow. 1987. Electron microscopical and serological detection of virus-like particles with lettuce big vein disease. J. Phytopathology 120:53–59.

von der Pahlen, A., and J. Crnko. 1965. El virus del mosaico de la lechuga (*Marmor lactucae*, Holmes) en Mendoza y Buenos Aires. Rev. Invest. Agropecuarias Ser. 5, 2(4):25–31.

Walkey, D. G. A., and D. A. C. Pink. 1990. Studies on resistance to beet western yellows virus in lettuce (*Lactuca sativa*) and the occurrence of field sources of the virus. Plant Pathology 39:141–155.

Walkey, D. G. A., and C. J. Bolland. 1987. Beet western yellows virus. Natl. Vegetable Res. Stn. Ann. Rpt. 1986/87:50–51.

Washek, R. L. 1983. Cucumber mosaic resistance in summer squash (*Cucurbita pepo* L.). Cornell Univ. PhD Thesis, Ithaca, NY.

Watts, J. E. 1975. The response of various breeding lines of lettuce to beet western yellow virus. Ann. Appl. Biol. 81:393–397.

Yuen, J. E., and J. W. Lorbeer. 1983. A new gene for resistance to *Bremia lactucae*. Phytopathology 73:159–162.

Zink, F. W., and J. E. Duffus. 1969. Relationships of turnip mosaic susceptibility and downy mildew (*Bremia lactuca*) resistance in lettuce. J. Amer. Soc. Hort. Sci. 94:403–407.

Zink, F. W., J. E. Duffus, and K. A. Kimble. 1973. Relationship of a nonlethal reaction to a virulent isolate of lettuce mosaic virus and turnip mosaic susceptibility in lettuce. J. Amer. Soc. Hort. Sci. 98:41–45.

Zitter, T. A., and V. L. Guzman. 1974. Incidence of lettuce mosaic and bidens mottle viruses in lettuce and escarole fields in Florida. Plant Dis. Reptr. 58:1087–1091.

Zitter, T. A., and V. L. Guzman. 1977. Evaluation of cos lettuce crosses, endive cultivars, and *Cichorium* introductions for resistance to bidens mottle virus. Plant Dis. Reptr. 61:767–770.

CHAPTER 4

Development and Breeding of Resistance to Pepper and Tomato Viruses

JON C. WATTERSON

Introduction

Solanaceous vegetables account for a large proportion of the human diet worldwide. Viral diseases continue to reduce potential yields despite even the most rigorous sanitation and insect control programs. Recent information (Tomlinson 1987) indicates that viral diseases of tomato and pepper are the most important viral diseases of vegetables in 17 of 19 countries under protected production and in 14 of 24 countries where crops are grown in the open field. The inability to adequately control viral diseases through crop management practices or by controlling insect vectors has led to an increased emphasis on disease resistance as the most effective method of crop protection. *Resistance* as used here is the ability of a plant to withstand a pathogen without sustaining damage. The economic advantages to farmers of planting new high-yielding varieties with genetic resistance to viruses are obvious.

Viral Diseases of Peppers

Many of the same viruses that infect solanaceous relatives such as tomato, eggplant, and tobacco are also pathogenic to pepper. Over 30

Dr. Watterson is Director, Plant Pathology at the Petoseed Company, Woodland, California 95695.

viruses are known to cause economic damage on pepper. Progress in breeding for resistance has been limited in the past to work on tobacco mosaic virus (TMV). In the last 10 years, however, breeders have begun developing resistance to potato virus Y (PVY), tobacco etch virus (TEV), pepper mottle virus (PeMV), pepper veinal mottle virus (PeVMV), and cucumber mosaic virus (CMV).

Tobacco Mosaic Virus (TMV)

TMV is the type member of the tobamovirus group. Viruses in this group are RNA-containing rigid particles approximately 18 × 300 nm. Over 200 species of plants are infected with this easily mechanically transmitted virus (Zaitlin and Israel 1975).

Holmes (1937) first described the *L* gene for resistance in *Capsicum frutescens* as a localizing (hypersensitive) reaction to TMV. Strains of the virus that will infect *L* genotypes have been known for some time (McKinney 1952). Tobamovirus strains that will attack new resistance genes in pepper have been continually described since 1977 (Boukema 1977). These pepper strains are serologically and biologically distinct from both tobacco mosaic and tomato mosaic viruses (Tobias et al. 1982; Wetter 1984). Names that have been used to describe these strains are samsun latent strain of tobacco mosaic virus, pepper unusual strain of tobacco mosaic virus, bell pepper mottle virus, pepper mild mottle virus and *Capsicum* mosaic virus (Pares 1985; Wetter and Conti 1988). The proposed name "pepper mild mottle virus" can easily be confused with mild isolates of pepper mottle virus, a potyvirus. Therefore, *Capsicum* mosaic virus, a more descriptive name, is used here for these biologically and serologically distinct pepper strains of TMV.

Four alleles for resistance to pathotypes of *Capsicum* mosaic virus have been described by Boukema (1984) and are presented in Table 1. The L^1 allele is common in many pepper varieties, and resistance appears to be inherited as a single dominant gene. Plants that are heterozygous (L^1/L^+) respond with a resistant reaction if subjected to moderate virus titers. If, however, heterozygous plants are grown under suboptimum conditions and inoculated with high virus titers, systemic necrosis and death may result. The response therefore is dependent on the host genotype and environmental conditions as well as virus titer.

Table 1. *Relation between genotypes for resistance in* Capsicum *and pathotypes of tobamovirus (after Boukema 1984; used with permission).*

		TOBAMOVIRUS PATHOTYPES			
CAPSICUM CV. OR ACCESSION	RESISTANCE GENOTYPE	P_0	P_1	$P_{1.2}$	$P_{1.2.3}$
C. annuum 'Early Calwonder'	L^+L^+	+	+	+	+
C. annuum 'Bruinsma Wonder'	L^1L^1	−	+	+	+
C. frutescens 'Tabasco'	L^2L^2	−	−	+	+
C. chinense PI 159236	L^3L^3	−	−	−	+
C. chacoense PI 260429 and SA 185	L^4L^4	−	−	−	−

+ = Systemic mosaic; susceptible.
− = Local lesions on inoculated leaves, no systemic mosaic; resistant.

Breeding for *Capsicum* mosaic virus resistance in peppers was given encouragement with the discovery of the L^4 gene in *C. chacoense* PI 260429. This optimism should be tempered, however, by the recognition that historically new strains of the virus constantly are selected by new *Capsicum* genes (Betti et al. 1986).

Pepper Potyviruses

The potyviruses are the largest and most important group of plant viruses. They are RNA-containing flexuous particles of approximately 11 × 730 nm and are characteristically aphid transmitted in a nonpersistent manner as well as mechanically transmissible (de Bokx and Huttinga 1981). Potato virus Y (PVY), tobacco etch virus (TEV), pepper mottle virus (PeMV), and pepper veinal mottle virus (PeVMV) are all members of this group and are infectious on peppers.

Considerable confusion surrounds the taxonomy of the potyviruses, and a good argument could be made for considering these viruses to be merely different strains of PVY. Comparative coat protein sequence data from PeMV and the tobacco veinal necrosis strain of PVY support the contention that these are both strains of PVY (van der Vlugt et al. 1989). Selassie et al. (1985) found only varying strains of PVY infecting peppers in France (Table 2). Makkouk and Gumpf (1976) reported considerable biological as well as serological variability among their California isolates, but they still considered them all to be PVY. New potyviruses affecting peppers continue to be described based primarily on serological tests (Fernandez-Northcote and Fulton 1980). The relationships between these new viruses and existing pepper host differentials remain to be determined. Considering the recognized plasticity within the potyviruses, it would appear that new pepper viruses and novel isolates in this group will be reported in the future.

Table 2. *Reactions to PVY strains on* Solanum tuberosum *and selected pepper varieties; characterization of pathotypes (after Selassie et al. 1985; used with permission).*

DIFFERENTIAL HOSTS	GROUP	PVY^O PVY^N	To-72	SON-41	Pi-72	Israel -82	VR-2	N17-E
Solanum tuberosum 'Bintje'		M	R	R	R	R	R	R
Capsicum annuum 'Anaheim F6'	1a	R	LL/NM	LL/NM	LL/NM	LL/NM	LL/NM	M
'Bastidon'	1a	R	N	N	M	M	N	M
'Yolo Wonder'	1b	R	M	M	M	M	M	M
'Yolo Y'	2	R	R	M	M	M	M	M
'Florida VR-2'	3	R	R	R	R	R	M	M
'Serrano V.C.'	4	R	R	R	R	R	R	R
PATHOTYPES		PVY-0			PVY-1		PVY-1-2	

Symbols: LL = local lesions; N = systemic vein necrosis; M = vein banding, 10 to 15 days after inoculation; R = resistant.

Severe leaf mottling and fruit distortion are symptoms caused by potato virus Y on pepper. This type member of the potyvirus group is worldwide in distribution. Simmonds and Harrison (1959) found that two recessive genes v_1 and v_2 controlled resistance to PVY in pepper. The symbol y^a was proposed (Cook and Anderson 1960) for the single recessive gene that conferred resistance to PVY derived from the pepper lines P11 and SC 46252. Later, Cook (1960) proposed changing this symbol to ey^a to designate a single recessive pleiotropic gene conditioning resistance to both PVY and TEV. However,

Cook (1961) found a single plant in the cultivar Yolo Wonder that was resistant to PVY but susceptible to TEV. This line was designated 'Yolo Y'. Thus, the symbol ey^a was considered invalid.

In later studies, Cook (1963) could not determine the exact gene linkages, but proposed that three loci were responsible for the responses to three different PVY strains. Resistance was inherited as single recessive alleles. Nagai (1968), working with the PVY strains present in Brazil, concluded that resistance at the *y* locus in pepper was conditioned by multiple recessive alleles, each with its own activity. Pochard (1976) proposed the alleles *poty*, which conferred resistance to both PVY and TEV, and *H*, which conferred only PVY resistance. Inheritance of resistance conferred by these alleles depended on the challenging PVY strains. "Partial dominance" to PVY has also been described (Shifriss and Marco 1980) for pepper. Plants in the heterozygous $Y^a y^a$ condition were differentiated in a repeated backcrossing program and considered partially resistant to the virus. Horvath (1986) reviewed the work in peppers and concluded that resistance was "inherited by polygenes."

It is evident that much confusion and seemingly contradictory results remain regarding the inheritance of PVY resistance in pepper. It is also apparent that in order to clarify the relationships between the host pepper genes and PVY strains we must establish (1) a standardized condition for growing plants; (2) a standardized method of inoculation; (3) uniformly reacting homozygous differential pepper lines; and (4) characterized pure isolates of the virus. Pathotype or strain classification schemes have been proposed by various pepper workers (Selassie et al. 1985; Pochard 1976; Makkouk and Gumpf 1974; and Nagai and Smith 1968). In order to serve the agricultural community, an international effort is now required in which pure characterized strains of PVY can be used in the classification of pepper resistance genes.

A second member of the potyvirus group, tobacco etch virus (TEV), is commonly found on peppers throughout the western hemisphere. Greenleaf (1956) found resistance to TEV in *Capsicum frutescens* (later changed to *C. chinense*) PI 152225 and *C. annuum* SC 46252 to be inherited as single recessive genes with one or more modifying genes. The symbols et^f and et^a were given to these genes though no studies were reported on their allelic nature. By using two

TEV isolates, Sowell and Demski (1977) confirmed the high levels of resistance in four plant introductions, but found others to be variable in reaction to viral inoculation. The resistance reported in the pepper varieties 'Florida VR-2', 'Delray', 'Avelar', 'Agronomico 8', and 'Agronomico 10' appears to be inherited as a recessive gene with a number of modifying genes that determine the intensity of the resistant response. The number of modifying genes present may be influenced by the susceptible parent (J. C. Watterson, unpublished) as well as by the TEV-resistant parent (Greenleaf 1956). Smith (1970) and later Makkouk and Gumpf (1974) proposed a strain-classification scheme whereby six pepper cultivars were used to differentiate five TEV strains. Smith further postulated that a pepper with resistance to TEV would also be resistant to PVY. As is the case for PVY, there is also a great need for an internationally recognized scheme of TEV strain–pepper gene identification based on differential host responses.

Pepper mottle virus (PeMV) is a serologically distinct member of the potyvirus group reported only in North America. Though it shares many biological properties with both TEV and PVY, PeMV can be distinguished from these latter two viruses by the local lesions and systemic necrosis produced on Tabasco pepper (Nelson et al. 1982). Tolerance to PeMV was found in 'Avelar' and 'Agronomico 8'. A single recessive gene, *pmv*, which had been reported to confer tolerance, was transferred to produce the variety 'Delray' (Zitter and Cook 1973). Since 'Avelar' was also resistant to PVY and TEV, Greenleaf (1986) proposed that the allele et^{av} be used to designate a gene with "higher potency than et^{a} which protects only against TEV-C (common strain) and PVY-N^{YR}". The highly resistant accessions from *Capsicum chinense* PI 152225 and PI 159236 are reported to contain recessive genes that are thought to be either allelic or located at two closely linked loci (Subramanya 1982). Nelson and Wheeler (1978) proposed a host differential series for strain separation.

In Africa and Asia, a fourth potyvirus, pepper veinal mottle virus (PeVMV), causes leaf mottling, leaf distortion, vein chlorosis, fruit distortion, and stunting on peppers. Though biologically similar to other viruses in the potyvirus group, it is serologically distinct (Brunt and Kenten 1972). Resistance to this virus was found in local peppers from Thailand and was determined to be inherited as a single recessive gene (Soh et al. 1977). Additive genes also appeared to play a role in the full expression of resistance.

Cucumber Mosaic Virus (CMV)

Cucumber mosaic virus is worldwide in distribution and affects more than 700 species of weeds and crop plants. It is a spherical 28 nm particle containing up to five distinct single-stranded RNA components, although only RNAs 1, 2, and 3 are required for infectivity (Francki et al. 1979). Resistance work has been hampered by (1) the discovery of few resistant *Capsicum* accessions; (2) the sensitivity of tolerant lines to methods of inoculation, host age, virus strains, and environmental conditions; (3) the association of poor horticultural characteristics with tolerance; and (4) the inability to stabilize the tolerant reaction (mild systemic symptoms) in homozygous breeding lines.

Early workers identified plant introductions as having tolerance, and Barrios et al. (1971) reported a high degree of tolerance in the accession LP-1. This tolerance was thought to be conditioned by a single recessive gene. Pochard (1976) found that over a period of three seasons in the field, the peppers 'Perennial' and *Capsicum baccatum* 3–4 appeared to have excellent tolerance. Later, 'Philomèle' and 'Zephyr' were reported by Pochard to be CMV-tolerant. Field evaluations of these lines have not been reported. Though other researchers have confirmed the tolerance of LP-1 'Perennial' and *C. baccatum*, no commercial variety with CMV resistance has been developed to date.

Viral Diseases of Tomato

Tomato is reported to be susceptible to over 40 viruses worldwide. However, because of its infectiousness and its potential to reduce yields up to 23% (Rast 1975), the tobamovirus tomato mosaic virus (ToMV) accounts for most of the effort to develop virus resistance in tomato. Depending on crop management, environmental conditions, insect vectors, and reservoir hosts, other viruses can also cause economic damage to tomato. Resistance breeding has been carried forward with cucumber mosaic virus (CMV), potato virus Y (PVY), tobacco etch virus (TEV), tomato yellow leaf curl virus (TYLC), beet curly top virus (BCTV), and tomato spotted wilt virus (TSWV).

Tomato Mosaic Virus (ToMV)

Tomato mosaic is by far the most common viral disease of tomato. It is extremely persistent and easily transmitted mechanically. It can contaminate soil as well as infect seed. ToMV is most closely related to tobacco mosaic virus (TMV) in the tobamovirus group, but can be differentiated by serological reactions, amino acid sequences of the coat protein, and host response. Typically, TMV infects systemically white burley tobacco, *Nicotiana sylvestris*, and *Datura stramonium*, whereas ToMV causes local lesions (Hollings and Huttinga 1976). ToMV symptoms on tomato may appear as a light and dark green mosaic, bright yellow mottling, a fern leaf growth, or necrosis of leaves, stems, and fruit. Symptoms, however, will vary not only according to the strain, but also depending on the light intensity, temperature, day length, age of the plant, and host genotype.

Breeding for resistance to this virus began in the 1940s. Walter and others had been using the recessive gene *Tm-1* in their programs. Not until Alexander released breeding lines with the $Tm2^2$ gene in 1963, however, did activity begin in earnest. Today, over 60 tomato lines and hybrids are grown with this genetic resistance. A proposal for the host gene and virus-pathotype relationship was presented by Pelham (1969). Rast (1975) found an additional pathotype, and now five virus strain groups are separated by their reactions on tomato lines with three different genes (Table 3). This classification scheme allows for rapid pathotype assignment and ease of handling for a breeding program. These pathotypes do not coincide, however, with antigenic relationships, coat-protein make-up, cross-protection characteristics, or reactions on tobacco.

Though susceptible to some strains, the *Tm-1* gene has been useful because of the prevalence in nature of strain 0 to which it is resistant. *Tm-2* for many years was linked to the virescent gene *nv* and, therefore, had limited value. This tight linkage has now been broken, thus making this gene more useful to plant breeders. The gene $Tm\text{-}2^2$ is widely used today and still remains effective against ToMV in commercial plantings worldwide. An isolate reported to overcome this resistance has not been found outside experimental greenhouses.

Manipulation of temperature and host age allow for the separation of susceptible, heterozygous resistant, and homozygous resistant individuals.

Table 3. *Relationships between genes for resistance in tomato with strains of tomato mosaic virus.*

	TOMATO MOSAIC VIRUS STRAINS				
PLANT GENOTYPE	0	1	2	1.2	2^2
(+/+)	S	S	S	S	S
TM-1/TM-1	T	S	T	S	R
TM-2/TM-2	R	R	S	S	R
TM-2^2/TM-2^2	R	R	R	R	S

S = Susceptible; T = Tolerant; R = Resistant.

Geminiviruses

The geminiviruses have characteristic paired (geminate) 15–22 nm icosahedral virus particles, each containing circular single-stranded DNA. The two most important groups of geminiviruses of tomato are beet curly top virus (BCTV), transmitted by the beet leafhopper (*Circulifer tenellus*), and the viruses causing tomato yellow leaf curl (TYLC), tomato leaf curl, tobacco leaf curl, chino del tomate, tomato yellow dwarf, and tomato golden mosaic, transmitted by the sweet potato whitefly (*Bemisia tabaci*)(Brown and Nelson 1988; Harrison 1985). The seemingly sudden appearance of *Bemisia* in new geographic areas as well as population explosions in established areas led to complete crop failures for some tomato producers in Mexico and Florida during the 1989–1990 winter growing season. Though little substantive information is available to explain these sudden epidemiological changes, nevertheless the following have undoubtedly had some impact: (1) the increases in long-distance transport, both within and between countries, of greenhouse-grown ornamentals and vegetable transplants; (2) the ineffectiveness of insecticides to control the whitefly vector; (3) the loss of beneficial predators through intensive insecticide programs; (4) the expansion of both acreage and growing seasons in mild-winter production regions.

Tomato yellow leaf curl (TYLC) is the most serious disease of tomatoes in the Middle East and North Africa (Czosnek et al. 1988; Ioannou 1985). Infected plants become stunted and develop small yellow curled leaflets. Fruit production stops abruptly as flowers abort. A level of TYLC tolerance expressed as mild symptoms was first found in *Lycopersicon pimpinellifolium* LA 121. This tolerance was inherited as a single, incompletely dominant gene (Pilowsky and Cohen 1974) and has been introduced into experimental hybrids. Better resistance has now been found in *L. cheesmanii*, *L. chilense*, *L. hirsutum*, and *L. peruvianum* (Kasrawi et al. 1988). These new sources of resistance from wild species appear to offer great potential if problems of hybrid fertility, linkage, and low recovery of resistance can be overcome.

Developing tomatoes with resistance to beet curly top virus (BCTV) has been for 50 years the main objective of the USDA tomato breeding program in Utah and later Washington. The virus is not mechanically transmitted, and progress has been slow because of the difficulty of dealing with the leafhopper vector and the complex resistance expressed in wild tomato species. Nevertheless, Martin and Thomas (1986) have released four tomato varieties from this program with increased levels of tolerance to establishment of infection and leafhopper preference. Inheritance of the resistance appears to be multigenic. Resistance is intermediate in a resistant × susceptible cross.

Cucumber Mosaic Virus (CMV)

In temperate regions, particularly where there are high concentrations of ornamentals, CMV can be a serious problem alone or, more frequently, in combination with other viruses. In 1972 and again in 1988, epiphytotics caused by a necrotic strain of CMV resulted in severe crop damage in France and Italy. A small satellite RNA in association with normal CMV strains was later found to be responsible for the lethal necrosis (Kaper and Waterworth 1977). Other satellite RNAs have since been shown to induce chlorosis or mild symptoms on tomato when inoculated along with CMV (Kurath and Palukaitis 1989).

Resistance to this virus has been reported for wild tomato relatives, including *L. cheesmanii*, *L. chmiewlewskii*, *L. hirsutum*, *L. parviflorum*, *L. peruvianum*, *L. penellii*, and *Solanum lycoper-*

sicoides. Kuriyama et al. (1971) found *L. peruvianum* PI 128648–6 to be resistant to CMV. Latterot (1980) later crossed this accession with PI 126926 and made numerous intercrossings to select the line 'CMVR *L. peruvianum*'. Phills et al. (1977) identified *Solanum lycopersicoides* PI 365378 as having good tolerance to CMV symptoms, but high titers of the virus could be recovered. *S. lycopersicoides* × *L. esculentum* hybrids also appeared to be highly tolerant in the greenhouse and in the field. Because of problems with infertility, definitive studies on the inheritance of resistance have not been reported. No tomato varieties have as yet been released with CMV resistance.

Tomato Potyviruses

Vein banding mosaic caused by potato virus Y (PVY) is aphid transmitted, as are all other members of the potyvirus group. PVY is most damaging in tropical and subtropical growing areas particularly when it is found in combination with other viruses such as TEV and ToMV. Beginning in 1955, Walter (1967) evaluated numerous *L. esculentum* accessions and found resistance. Unfortunately, each time he found resistance, a new PVY strain appeared. He is credited with discovering four new tomato pathotypes of PVY. Since then, recessively inherited resistance in *L. hirsutum* PI 247087 and PI 365903 has been reported (Anonymous 1986; Thomas and McGrath 1988). The South American variety 'Angela' also has resistance to PVY (Watterson 1986).

Tobacco etch virus (TEV) causes plant stunting and mottling along with leaf distortion. Holmes (1946) found *L. chilense* and *L. hirsutum* to be good sources of resistance. According to Walter (1967), *L. esculentum* PI accessions 183692 and 166989 were resistant, and inheritance was conditioned by a single recessive gene. Breeding stocks with this resistance were developed but apparently never widely distributed. Because of recent outbreaks of this disease in the major tomato production areas of Mexico, re-evaluation of these breeding stocks seems warranted.

Tomato Spotted Wilt Virus (TSWV)

TSWV is a unique RNA virus 70 × 90 nm that possesses a membrane envelope surrounding the viral protein coat. It infects more than 300 plant species, including dicots and monocots (Cho et al. 1989). In the field, the virus is transmitted by adult thrips, though only the larvae can acquire the virus. The diversity of hosts of the virus, particularly in ornamentals, has resulted in more frequent occurrences of TSWV in nursery-grown tomato plants. Incidence of the tomato spotted wilt virus (TSWV) in tomatoes has increased steadily in North America over the last five years. Among possible reasons for the sudden increase in losses to this disease are (1) changes in the use of insecticides, particularly the synthetic pyrethroids; (2) the wider geographic distribution of the western flower thrips; and (3) the increasing use of greenhouse-grown tomato transplants, where the thrips' vectors are commonly present. Other tomato-producing regions of the world are also reporting the presence of this virus for the first time. Symptoms of the disease vary greatly, depending on the predominant strain or strains, but commonly consist of yellow and necrotic spotting of the leaves, which gives a bronze appearance to the foliage. Greasy or silvery streaks often develop on one side of stems or petioles. Affected plants are stunted. Ring spots on leaves and fruit are often quite prominent. Some strains cause a necrotic dieback or tip blight that results in the death of the plant.

The 50-year history of work to produce TSWV-resistant tomatoes is comparable in many ways to work with the Beet curly top virus. Tolerance to TSWV has been known for some time; however, the development of commercially successful tomato varieties that have good horticultural characteristics and consistent performance over wide geographic areas has not been forthcoming. The reasons for this poor record are understandable. Because of the unstable nature of the virus in crude sap, even under optimum conditions mechanical transmission may only approach 80%. In addition, numerous strains of the virus are known to occur, and resistance appears to be strain-specific. Initial observations of apparent single-gene resistance are now confounded by modifying gene action (Watterson et al. 1989). Wenholz (1939) reported that *L. pimpinellifolium* and a "Peruvian type" had good resistance to TSWV. Kikuta and Frazier (1946) used a *L. esculentum* × *L. pimpinellifolium* selection as the source of resistance for their variety 'Pearl Harbor'. Researchers in Australia, Argentina, California, and New Jersey gave varying reports on resistance for 'Pearl Harbor' as well as 'Rey de Los Tempranos', 'Manzana' and 'German Sugar'. Finlay (1952), in Australia, proposed that the breakdown of resistance in different regions was due to the presence of different TSWV strains. He grouped isolates into ten distinct strains based on the type of symptoms produced and the different varietal reactions. Despite the presence of at least five genes for resistance, new tomato lines continued to develop TSWV symptoms. Finlay concluded that there were great "benefits to be gained by perseverance with hybrids involving *L. peruvianum* as a source of resistance." Smith (1944) also pointed out the excellent resistance present in a number of accessions of this latter species. Recent work in this laboratory and in Hawaii (Watterson et al. 1989) has led to the development of TSWV-resistant lines that derive their original resistance from *L. peruvianum*. Inheritance appears to be due to a major dominant gene with undetermined modifying genes. Resistant individual plant selections are not systemically infected, but produce a hypersensitive response limited to inoculation sites. Resistance appears to be effective against all tested strains.

Breeding Virus-Resistant Vegetables: Practical Approaches

Three ingredients are essential in a successful virus-resistance breeding program: (1) there must be heritable resistance present in the plant species; (2) virus isolates must be pure as well as representative of field isolates; and (3) an accurate and highly efficient method of screening segregating populations should be established.

With some viruses, determining resistance in the host can be difficult. For example, the transmission rate for TSWV in tomatoes inoculated in the greenhouse is reported to be 50–80%. However, some modifications can be used to improve upon this rate. Plants are inoculated in the greenhouse at the fifth leaf stage, which is the optimum age for transmission. Six leaves are inoculated at any one inoculation. Sodium sulfite, an antioxidant, is added to a cold phosphate buffer and used to dilute TSWV-infected plant sap. A typical TSWV-breeding scheme is presented in Table 4. Plants that do not develop either a typically localized susceptible or resistant reaction are reinoculated one week later. Resistant selections are transplanted to the field, where late-expressing susceptible individuals are eliminated. While crosses are being made, self-pollinated seed is also collected from plants that remain healthy in the field. Small populations of this next generation are then challenged with the virus. Any individual selections that continue to segregate for resistance are discarded. In this way, the fewest number of escapes are carried forward in the breeding programs.

Table 4. *Petoseed/Hawaii TSWV Resistance Program.*

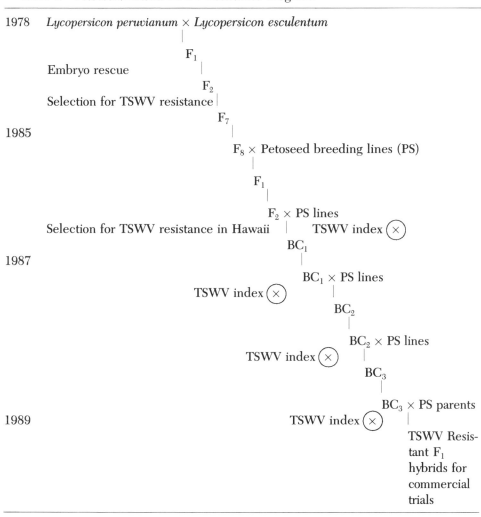

1978	*Lycopersicon peruvianum* × *Lycopersicon esculentum*
	F_1
	Embryo rescue
	F_2
	Selection for TSWV resistance
	F_7
1985	
	F_8 × Petoseed breeding lines (PS)
	F_1
	F_2 × PS lines
	Selection for TSWV resistance in Hawaii; TSWV index (×)
	BC_1
1987	
	BC_1 × PS lines
	TSWV index (×)
	BC_2
	BC_2 × PS lines
	TSWV index (×)
	BC_3
	BC_3 × PS parents
1989	TSWV index (×)
	TSWV Resistant F_1 hybrids for commercial trials

In host-pathogen systems in which a single dominant gene conditions resistance, such as ToMV of tomato, the procedures and breeding schemes are often simpler. ToMV-resistant individuals are identified in a seedling stage by a hypersensitive response and used in backcrossing to a horticulturally desirable recurrent parent. When horticulturally superior resistant individuals are identified, often 5–7 generations later, a final generation of selfing is required. Small populations of 20–30 plants from each individual plant selection are challenged with ToMV in a progeny test of the selfed generation. Selections identified as either homozygous susceptible or heterozygous are discarded. The remaining homozygous resistant individuals can then be bulked.

The apparent simplicity of such a scheme for selecting plants resistant to a virus can, however, be misleading. Considerable research is required in advance to select, establish, and maintain pure isolates of the virus. Viruses such as ToMV or *Capsicum* mosaic virus of pepper have many distinct strains that can be recognized serologically and with host differentials (Tobias et al. 1982). The isolates used in a virus screen must coincide with the resistance genes present in the host. For mechanically transmitted viruses, care must be taken to prevent contamination from undesirable viruses that may be seed-borne (e.g. ToMV and *Capsicum* mosaic virus), carried by insect vectors, transmitted by workers, or spread through poor sanitation practices. The purity of established virus isolates must be assured by selective host passage, sequential single lesion transfer, or storage in dessicated plant material. Easily identifiable strains causing ringspots, necrosis, or white mosaic patterns can be used to efficiently screen large plant populations. When such strains are used, susceptible individuals can quickly be recognized and discarded from a breeding program.

Environmental factors must be established in artificial inoculations that allow for distinct reactions between resistant and susceptible individuals. Temperature and light are often critical. For example, under conditions of low light intensity, pepper plants susceptible to PVY will remain symptomless until higher light conditions return. Warm temperatures are usually favorable for viruses of pepper and tomato; however, by changing temperatures from 28°C to 20°C after inoculating tomatoes with ToMV, a necrotic reaction can be induced on heterozygous plants (Alexander and Cirulli 1966). This temperature shift allows the pathologist or breeder to identify the genotypic makeup of individual tomato seedlings. The result is that 75% of the plants in the F_2 population can be eliminated, while the remaining 25% homozygous resistant individuals are carried on.

After manipulation of the host, virus, and the environmental factors, a disease screen is established. The procedures used must be appropriate to the breeding objectives. With single gene resistances, small populations can be used and examined very intensively. Multigenic resistance or major genes with additive gene action, however, are more often the rule. In these cases, a practical program for inoculating thousands of individuals must be established. In our own program for developing TEV-resistant peppers, more than 100,000 plants may be inoculated in a single test. The scale of such an operation requires efficient inoculation techniques and controlled environmental conditions. The inclusion of proper differential hosts as controls is particularly important for such a virus screen. Reactions of control plants can then be used to estimate the numbers of escapes in the test. The breeder can compensate for these escapes by over-selecting within his breeding material. Progeny from single plant selections are evaluated for resistance in the next generation, and any susceptible lines are discarded. The object of these screens is to identify resistant or tolerant plants while eliminating as many escapes as possible. Valuable time of both pathologist and breeder can be wasted on individuals which later turn out to be escapes.

Breeding Virus-Resistant Vegetables: Future Possibilities

What does the future hold for us? With the new technological advances made in gene transfer and transformation of plants, the introduction of foreign genes to control viral diseases is possible. Virus coat protein genes for ToMV have now been introduced into tomato to provide cross protection against ToMV (Nelson et al. 1988). Though the prospects are exciting, the efficacy under conditions of natural infection remains to be demonstrated. The introduction of cDNA copies of satellite RNA such as disease-ameliorating isolates of CMV CARNA 5 into the plant genome offers another possible mechanism for protecting vegetables against viruses (Palukaitis 1988). Developing selection systems for resistance expressed at a cellular level would be a major step in the advancement of breeding for virus resistance in vegetables. New highly sensitive methods for quantifying virus concentration such as ELISA, cDNA, and dot blot hybridization may in the future allow a rapid and more precise assessment of viral resistance. Although the new technologies will unquestionably play an important role in virus resistance work, we should not forget the untapped treasures that still exist in the wild gene pools. We have not yet exploited the full potential of presently available germplasm. We should be alarmed by the destruction and loss of the wild species germplasm. How many accessions of *Capsicum chacoense*, the source of *Capsicum* mosaic virus resistance, do we have in the U.S. collections? How homogeneous is this collection? What is an adequate sample of the population? Does it fully represent the geographic distribution of this species? What other pest or stress-tolerance genes are present but unselected? These are questions that should be answered for all crop species.

In our attempts to increase the variability of our crops by cellular and genetic manipulation, it is critical that we not neglect the potentially valuable resources available in wild relatives of crop species. Whether it is cold tolerance, insect tolerance, ozone resistance, salt tolerance, virus resistance, or other characteristics, much remains locked in the plant genome. Considerable effort devoted to the collection, evaluation, and maintenance of germplasm will be necessary to provide the keys to unlock these secrets. Finally, great diligence will be required to apply these basic discoveries for the improvement of our vegetable crops.

Literature Cited

Alexander, L. J., and M. Cirulli. 1966. Inheritance of resistance to tobacco mosaic virus in tomato. Phytopathology 56:869.

Anonymous. 1986. AVRDC Progr. Rep. Summaries. p. 45.

Barrios, E. P., H. I. Mosakar, and L. L. Black. 1971. Inheritance of resistance to tobacco and cucumber mosaic viruses in *Capsicum frutescens.* Phytopathology 61:1318.

Betti, L., M. Tanzi, and A. Canova. 1986. Increased pathogenicity of TMV pepper strains after repeated passages in resistant *Capsicum* accessions. *Capsicum* Newslett. 5:43–44.

Boukema, I. W. 1984. Resistance to TMV in *Capsicum chacoense* Hunz is governed by an allele of the *L*-locus. *Capsicum* Newslett. 3:47–48.

Brown, J. K. and M. R. Nelson. 1988. Transmission, host range, and virus-vector relationships of chino del tomate virus, a whitefly-transmitted geminivirus from Sinaloa, Mexico. Plant Dis. 72:866–869.

Brunt, A. A., and R. H. Kenten. 1972. Pepper veinal mottle virus. Descr. Plant Viruses No. 104. Kew, Surrey, England: Commonw. Mycol. Inst./Assn. Appl. Biol.

Cho, J. J., R. F. L. Mau, T. L. German, R. W. Hartman, L. S. Yudin, D. Gonsalves, and R. Provvidenti. 1989. A multidisciplinary approach to management of tomato spotted wilt virus in Hawaii. Plant Dis. 73:375–383.

Cook. A. A. 1960. Genetics of resistance in *Capsicum annuum* to two virus diseases. Phytopathology 50:364–367.

Cook, A. A. 1961. A mutation for resistance to potato virus Y in pepper. Phytopathology 51:550–552.

Cook, A. A., 1963. Genetics of response in pepper to three strains of potato virus Y. Phytopathology 53:720–722.

Cook, A. A., and C. W. Anderson. 1960. Inheritance of resistance to potato virus Y derived from two strains of *Capsicum annuum.* Phytopathology 50:73–75.

Czosnek, H., R. Ber, Y. Antignus, S. Cohen, N. Navot, and D. Zamir. 1988. Isolation of tomato yellow leaf curl virus, a geminivirus. Phytopathology 78:508–512.

de Bokx, J. A., and H. Huttinga. 1981. Potato virus Y. Descr. Plant Viruses No. 242. Kew, Surrey, England: Commonw. Mycol. Inst./Assn. Appl. Biol.

Fernandez-Northcote, E. N., and R. W. Fulton. 1980. Detection and characterization of Peru tomato virus strains infecting pepper and tomato in Peru. Phytopathology 70:315–320.

Finlay, K. W. 1952. Inheritance of spotted wilt resistance in the tomato. I. Identification of strains of the virus by the resistance or susceptibility of tomato species. Aust. J. Agr. Res. B 5:303–314.

Francki, R. I. B., D. W. Mossop, and T. Hatta. 1979. Cucumber mosaic virus. Assn. Appl. Biol. Descr. Plant Viruses No. 213. Kew, Surrey, England: Commonw. Mycol. Inst. Assn. Appl. Biol.

Greenleaf, W. 1956. Inheritance of resistance to tobacco etch virus in *Capsicum frutescens* and in *Capsicum annuum*. Phytopathology 46:371–375.

Greenleaf, W. 1986. Pepper breeding. In Breeding Vegetable Crops. Ed. M. J. Bassett. Westport, CT: AVI Publishing Co., Inc.

Harrison, B. D. 1985. Advances in geminivirus research. Ann. Rev. Phytopathology 23:55–82.

Hollings, M., and H. Huttinga. 1976. Tomato mosaic virus. Descr. Plant Viruses No. 156. Kew, Surrey, England: Commonw. Mycol. Inst./Assn. Appl. Biol.

Holmes, F. O. 1937. Inheritance of resistance to tobacco mosaic disease in the pepper. Phytopathology 27:637–642.

Holmes, F. O. 1946. A comparison of the experimental host ranges of tobacco-etch and tobacco-mosaic viruses. Phytopathology 36:643–659.

Horvath, J. 1986. Compatible and incompatible relations between *Capsicum* species and viruses. I. Review. Acta Phytopathol. 21:35–49.

Ioannou, N. 1985. Yield losses and resistance of tomato strains of tomato yellow leaf curl and tobacco mosaic viruses. Agr. Res. Inst. (Cyprus) Tech. Bull. 66. 11 pp.

Kaper, J. M., and J. E. Waterworth. 1977. Cucumber mosaic virus associated RNA 5: Causal agent for tomato necrosis. Science 196:429–431.

Kasrawi, M. A., M. A. Surwan, and A. Mansour. 1988. Sources of resistance to tomato yellow leaf curl virus (TYLCV) in *Lycopersicon* species. Euphytica 37:61–64.

Kikuta, K., and W. A. Frazier. 1946. Breeding tomatoes for resistance to spotted wilt in Hawaii. Proc. Am. Soc. Hort. Sci 47:271–276.

Kurath, G., and P. Palukaitis. 1989. Satellite RNAs of cucumber mosaic virus: Recombinants constructed *in vitro* reveal independent functional domains for chlorosis and necrosis in tomato. Mol. Plant Microbe Interact. 2:91–96.

Kuriyama, T., K. Kuniyasu,and H. Mochizuki. 1971. Studies on the breeding of disease resistance tomatoes by interspecific hybridization. II. Fertility and disease resistance in the progenies of interspecific hybridization. Bull. Hort. Res. Sta. B. (Okitsu) 11:33–60.

Laterrot, H. 1980. Recherches sur la "tomate". In Rapport 1979–1980 de la station d'amélioration des plantes maraichères d'Avignon, 84140 Montfavet, France. pp. 110–112.

Makkouk, K., and D. Gumpf. 1974. Further identification of naturally occurring virus diseases of pepper in California. Plant Dis. Rep. 58:1002–1006.

Makkouk, K., and D. Gumpf. 1976. Characterization of potato virus Y strains isolated from pepper. Phytopathology 66:576–581.

Martin, M. W., and P. E. Thomas. 1986. Levels, dependability, and usefulness of resistance to tomato curly top disease. Plant Dis. 70:136–141.

McKinney, H. H. 1952. Two strains of tobacco mosaic virus, one of which is seed-borne in an etch immune pungent pepper. Plant Dis. Rep. 36:184–187.

Nagai, H. 1968. Obtencao de variedades de pimentào resistêntes ao mosaico Bragantia 17:311–354.

Nagai, H., and P. Smith. 1968. Reaction of pepper varieties to naturally occurring viruses in California. Plant Dis. Rep. 52:929–930.

Nelson, M. R., and T. A. Zitter. 1982. Pepper mottle virus. Descr. Plant Viruses No. 253. Kew, Surrey, England: Commonw. Mycol. Inst./Assn. Appl. Biol.

Nelson, M. R., and R. E. Wheeler, and T. A. Zitter. 1982. Pepper mottle virus. Descr. Plant Viruses No. 253. Kew, Surrey, England: Commonw. Mycol. Inst./Assn. Appl. Biol.

Nelson, R. S., S. M McCormick, X. Delannay, P. Dube, J. Layton, E. G. Anderson, M. Kaniewska, R. K. Proksh, R. B. Horsch, S. G. Rogers, R. T. Fraley, and R. N. Beachy. 1988. Virus tolerance, plant growth, and field performance of transgenic tomato plants expressing coat protein from tobacco mosaic virus. Bio/Technology 6:403–409.

Palukaitis, P. 1988. Pathogenicity regulation by satellite RNAs of cucumber mosaic virus: Minor nucleotide sequence changes alter host responses. Mol. Plant Microbe Interact. 1:175–181.

Pares, R. D. 1985. A tobamovirus infecting *Capsicum* in Australia. Ann. Appl. Biol. 106:469–474.

Pelham, J. 1969. Isogenic lines to identify physiologic strains of TMV. Tomato Genet. Coop. Rep. 19:18.

Phills, B. R., R. Provvidenti, and R. W. Robinson. 1977. Reaction of *Solanum lycopersicoides* to viral diseases of the tomato. Tomato Genet. Coop. Rep. 27:18.

Pilowsky, M., and S. Cohen. 1974. Inheritance of resistance to tomato yellow leaf curl virus in tomatoes. Phytopathology 64:632–635.

Pochard, E. 1976. Recherches sur le "piment". In Rapport 1975–1976 de la station d'amélioration des plantes maraîchères d'Avignon, 84140 Montfavet, France. pp. 52–54.

Rast, A. T. B. 1975. Variability of tobacco mosaic virus in relation to control of tomato mosaic in glasshouse tomato crops by resistance breeding and cross protection. Inst. Phytopathol. Res. Wageningen Publ. 689. 75 pp.

Selassie, K. G., G. Marchoux, B. Delecolle, and E. Pochard. 1985. Variabilité naturelle des souches du virus Y de la pomme de terre dans les cultures de piment du sud-est de la France. Caractérisation et classification en pathotypes. Agronomie 5:621–630.

Shifriss, C., and S. Marco. 1980. Partial dominance of resistance to potato virus Y in *Capsicum.* Plant Dis. 64:57–59.

Simmonds, N., and E. Harrison. 1959. The genetics of reaction to pepper vein-banding virus. Genetics 44:1281–1289.

Smith, P. G. 1944. Reaction of *Lycopersicon* spp. to spotted wilt. Phytopathology 34:504–505.

Smith, P. G. 1970. Tobacco etch strains on peppers. Plant Dis. Rep. 54:786–787.

Soh, A. C., T. C. Yap, and K. M. Graham. 1977. Inheritance of resistance to pepper veinal mottle virus in chili. Phytopathology 67:115–117.

Sowell, G., and J. W. Demski. 1977. Resistance of plant introductions of pepper to tobacco etch virus. Plant Dis. Rep. 61:146–148.

Subramanya, R. 1982. Relationship between tolerance and resistance to pepper mottle virus in a cross between *Capsicum annuum* L. and *Capsicum chinense* Jacq. Euphytica 31:461–464.

Thomas, J. E., and D. J. McGrath. 1988. Inheritance of resistance to potato virus Y in tomato. Austr. J. Agr. Res. 39:475–479.

Tobias, I., A. T. B. Rast, and D. Z. Maat. 1982. Tobamoviruses of pepper, eggplant and tobacco: Comparative host reactions and serological relationships. Netherlands J. Plant Pathol. 88:257–268.

Tomlinson, J. A. 1987. Epidemiology and control of virus diseases of vegetables. Ann. Appl. Biol. 110:661–681.

Van der Vlugt, R., S. Allefs, P. de Haan, and R. Goldbach. 1989. Nucleotide sequence of the 3′-terminal region of Potato virus y^N RNA. J. Gen. Virol. 70:229–233.

Walter, J. M. 1967. Hereditary resistance to disease in tomato. Ann. Rev. Phytopathology 5:131–162.

Watterson, J. C. 1986. Diseases. In The Tomato Crop. Eds. J. G. Atherton and J. Rudich. New York: Chapman and Hall.

Watterson, J. C., C. Wyatt, and J. Cho. 1989. Inheritance of tomato spotted wilt resistance in tomato. Int. Conf. on TSWV, March 27–29, 1989, Honolulu, Hawaii.

Wenholz, H. 1939. Spotted wilt of tomatoes. Breeding for resistance. Hawkesbury Agr. Coll. J. 36:103.

Wetter, C. 1984. Serological identification of four tobamoviruses infecting pepper. Plant Dis. 68:597–599.

Wetter, C., and M. Conti. 1988. Pepper mild mottle virus. Descr. Plant Viruses No. 330. Kew, Surrey, England: Commonw. Mycol. Inst./Assn. Appl. Biol.

Zaitlin, M., and H. W. Israel. 1975. Tobacco mosaic virus (type strain). Descr. Plant Viruses No. 151. Kew, Surrey, England: Commonw. Mycol. Inst./Assn. Appl. Biol.

Zitter, T., and A. Cook. 1973. Inheritance of tolerance to a pepper virus in Florida. Phytopathology 63:1211–1212.

CHAPTER 5

DNA Markers for Viral Resistance Genes in Tomato

NEVIN D. YOUNG, RAMON MESSEGUER,
DANIEL B. GOLEMBOSKI & STEVEN D. TANKSLEY

Introduction

Many examples of major plant genes conferring resistance to microbial pathogens are known. These genes have been extremely valuable in agriculture by providing effective protection from otherwise serious plant diseases. Genes for resistance to viral, bacterial, and fungal pathogens have been identified and used extensively in tomato-breeding programs. The inheritance of some of these resistance genes is known and, in many cases, these traits are controlled by a single dominant or incompletely dominant gene (Pelham 1966). It has even been possible to infer the general mechanism for viral resistance in some genes by determining whether resistance inhibits viral replication, as in the tobacco mosaic virus (TMV)-resistant gene,

Dr. Young is in the Department of Plant Pathology, University of Minnesota, St. Paul, MN 55108. Drs. Messeguer and Tanksley are in the Department of Plant Breeding, and Dr. Golemboski is in the Department of Plant Pathology, Cornell University, Ithaca, NY 14853. The authors thank M. Zaitlin, M. Sorrells, M. Ganal, M. Roeder, J. Miller, and W. Wu for their valuable comments and assistance. This work was supported by grant US-1388-87 from the US-Israel Binational Agricultural Research and Development Fund, 85-CRCR-1-1609; the United

Tm-1 (Fraser and Loughlin 1980), or cell-to-cell movement, as in the TMV resistance gene, *Tm-2* (Motoyoshi and Oshima 1977). Nevertheless, the molecular basis of these and other plant disease resistance genes is unknown and will probably remain so until the actual DNA sequences coding for resistance are cloned and analyzed.

Cloning disease-resistance genes in tomato (or any other higher plant) has proven difficult because little or nothing is known about the gene product(s) synthesized by the resistance genes. Current gene cloning techniques rely on prior knowledge of a gene's mRNA or protein product (Sambrook et al. 1989). In the absence of this knowledge, gene cloning has previously been impossible. For example, several attempts have been made to isolate the *N*-gene of tobacco (conferring a hypersensitive form of resistance to TMV) by differential screening of cDNA libraries probed with mRNA from induced versus uninduced leaves, but without success (Dunigan et al. 1987).

Map-based Cloning

Recently, a new approach to gene cloning, suitable for disease-resistance genes in plants, has been developed. This strategy is known as *map-based cloning* and relies not on knowledge of a gene's RNA or protein product, but on knowledge of its genetic map position (Orkin 1986; Tanksley et al. 1989). Map-based cloning begins with the identification of restriction fragment length polymorphisms (RFLPs) that are very tightly linked genetically to the gene of interest. RFLPs are cloned DNA sequences that can be used to construct a genetic linkage map by hybridization to Southern blots containing DNA from a group of related individuals, such as an F_2 or backcross population. Whenever the parents of the population differ at the DNA sequence

States Department of Agriculture; and the Cornell Biotechnology Program, which is sponsored by the New York Science and Technology Foundation, a consortium of industry, the US Army Research Office, and the National Science Foundation. This paper is published as contribution No. 18,052 of the series of the Minnesota Agricultural Experiment Station based on research conducted under Project 014, supported by General Agricultural Research funds.

level, the progeny segregate for differing allelic forms of the cloned DNA segment. If the population also segregates for a gene of interest (such as disease resistance), the genetic location of each RFLP marker also can be determined relative to that gene. Tightly linked RFLPs then act as starting points for chromosome walking (Steinmetz et al. 1981), in which nearby portions of the chromosome are isolated in a series of overlapping steps, thus eventually leading to the isolation of a genomic segment containing the target-resistance gene.

Two recent techniques have made such chromosome walking over long genomic distances feasible. One is known as pulsed field gel electrophoresis (Carle et al. 1986; Chu et al. 1986; Ganal and Tanksley 1989). This technique makes resolving large DNA fragments up to ten million base pairs in length possible. Thus, RFLP markers, along with the target gene, can be mapped on a physical (nucleotide base pairs) as well as genetic (recombinational) basis. The second technique involves cloning large DNA fragments, up to 300,000 base pairs or more in length, into yeast artificial chromosome vectors (Burke et al. 1987). This enables each step of a chromosome walk to proceed a sizeable distance so that relatively few total steps are required to move a significant distance along the chromosome. The final phase in map-based cloning of a disease-resistance gene consists of subcloning large DNA fragments containing the target gene into smaller, transformation-type vectors and testing the subclones by insertion into the genome of a sensitive plant host. Fortunately, producing these independent transformants is possible in tomato and many other plants (McCormick et al. 1986).

In our labs, we are developing tools to apply map-based cloning to several disease-resistance genes in tomato and other crop species. Two of the gene targets in tomato are single-locus genes, known as *Tm-1* and *Tm-2*, that confer resistance to TMV. Here, we describe progress in this effort.

Using Isolines to Idenitfy Tightly Linked RFLPs

The first step in map-based cloning is the identification of RFLP markers that are very tightly linked to the target-resistance gene. To be useful as starting points for chromosome walking, these RFLPs must be very close to the target gene, generally less than one recombination unit away, since only very tightly linked DNA markers are physically close enough to be useful. The process of finding such tightly linked RFLP markers would be difficult and extremely time-consuming, however, if each RFLP had to be screened one at a time and analyzed by complete segregation analysis. Therefore, we have developed an improved strategy for rapidly identifying tightly linked RFLP markers and have applied this strategy to find RFLPs near both the *Tm-1* and *Tm-2* genes of tomato (Young et al. 1988).

This strategy is based on the use of nearly isogenic lines (NILs), which have been developed for many disease-resistance genes in tomato. Both *Tm-1* and *Tm-2* were originally derived from wild relatives of cultivated tomato and introduced into susceptible cultivated lines by repeated backcross selection breeding. Backcross breeding begins with the production of a hybrid between the donor (wild) and recipient (cultivated) tomato lines. This F_1 hybrid is then backcrossed to the susceptible cultivated tomato parent and the progeny screened for resistance to the virus. This process is repeated several times until a line is obtained that carries the target-resistance gene in a background that is nearly identical to that of the cultivated tomato parent, except for the resistance gene and a small chromosomal region flanking it (known as the *introgressed segment*). Because the donor species of *Tm-1* and *Tm-2* are divergent compared to cultivated tomato in terms of nucleotide sequence, most genomic clones located in the introgressed segments can be identified by virtue of a DNA polymorphism between the NILs. Genomic clones located outside the introgressed regions normally show no polymorphism. In effect, the introgressed segment acts as a bull's eye for assaying whether a given RFLP is located in the tightly linked chromosomal segment flanking the *Tm-1* or *Tm-2* (Figure 1).

To speed further the process of identifying RFLPs tightly linked to *Tm-1* and *Tm-2*, we have also adopted the strategy of pooling groups of five genomic clones together for assaying with NILs (Young et al. 1988, 1987). Since only those clones that map within the introgressed segment are likely to be polymorphic, it is feasible to assay multiple clones simultaneously and still distinguish rare RFLPs that are tightly linked to the target genes.

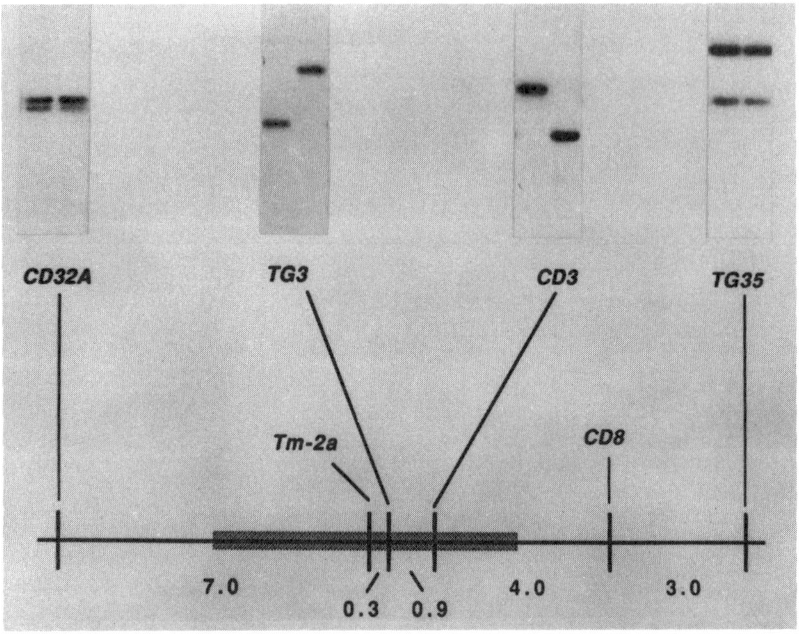

Figure 1. *RFLPs located in the introgressed region of NILs. At the* bottom, *the genetic map of chromosome 9 of tomato is shown. The* shaded region *indicates the introgressed segment between the NILs, New Yorker and New Yorker-*Tm2-2 *(kindly provided by R. Provvidenti, Geneva, NY). At the* top, *strips show the restriction fragment patterns of four DNA markers (with New Yorker on the* left *and New Yorker-*Tm2-2 *on the* right*). Note that only those DNA markers that map within the introgressed segment show different-sized bands on the strips. (1990,* Bio/Technology. *Used by permission.)*

RFLPs Near *Tm-1* and *Tm-2*

A single-copy genomic library of tomato genomic sequences was created by *Pst*I digestion of high-molecular-weight tomato DNA, size selection, insertion into the plasmid pUC-8, and screening and removal of organellar and repetitive sequences (Miller and Tanksley 1990). Groups of five clones from this library were then probed against Southern blots of restriction enzyme-digested DNA from pairs of NILs with or without the *Tm-1* or *Tm-2* gene (Figure 2). The NILs for *Tm-1* were the kind gift of H. Latterot, Montfavet, France;

the NILs for *Tm-2* were developed by Drs. H. M. Munger and M. A. Mutschler, Cornell University, Ithaca, NY. Out of 120 groups of pooled clones (600 total clones), 20 contained a polymorphism associated with the *Tm-1* gene and three contained a polymorphism associated with *Tm-2*. An example of one group showing a polymorphism between NILs with or without the *Tm-1* gene is shown in Figure 2. In several cases, the groups were subdivided into individual clones and tested against Southern blots of the appropriate NILs, thereby pinpointing the precise clone located near the gene of interest.

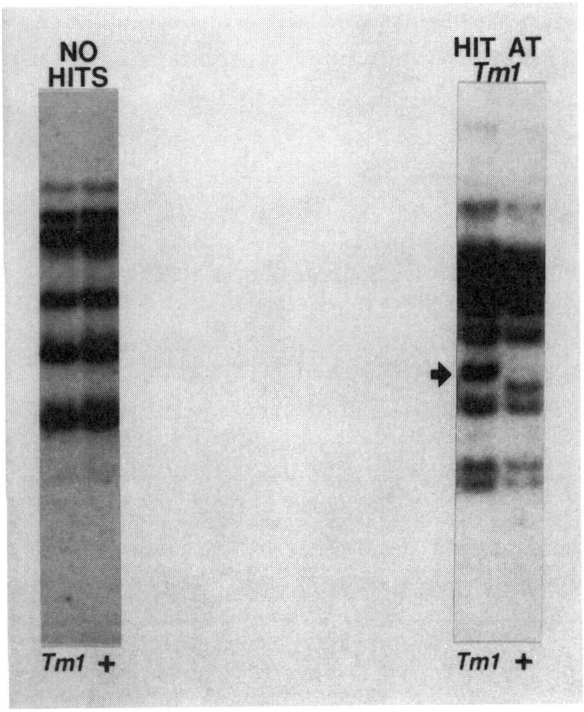

Figure 2. *Screening groups of tomato genomic clones for proximity to* Tm-1 *by use of NILs. A total of 120 groups, each consisting of five genomic clones pooled together, were radiolabeled and probed against Southern blots of DNA from a pair of NILs with or without* Tm-1. *The* left *portion of the figure illustrates one group without polymorphisms between the NILs; the* right *illustrates a group of clones exhibiting a polymorphism between the NILs. The band that is polymorphic is noted by an* arrow.

To confirm that these clones were tightly linked to the resistance genes, some of the clones were further analyzed in F_2 populations, segregating for either the *Tm-1* or *Tm-2* genes. As expected from previous results (Young et al. 1988), all clones putatively linked to the *Tm-2* gene were found to map at the same locus near the center of chromosome 9 at precisely the same position as *Tm-2*. RFLPs putatively linked to the *Tm-1* gene were found to map near the top of chromosome 2, near the *R45S* locus of tomato (Vallejos et al. 1986).

Simultaneously, we also probed additional sets of NILs that distinguish RFLPs near other important tomato genes conferring resistance to bacterial, fungal, and nematode pathogens (data not shown). Thus, in a single hybridization experiment, several different DNA clones could to be tested simultaneously for linkage to several different gene targets. This approach is clearly more efficient than targeting one resistance gene at a time and emphasizes the power of targeting genes by multiple probing of NILs.

Not only will RFLPs near tomato-disease-resistance genes be useful in map-based cloning, they may also be helpful in marker-assisted selection (Tanksley et al. 1989; Young and Tanksley 1989). The application of RFLPs to marker-assisted selection makes scoring for the presence of a disease-resistance allele possible without actual disease testing. Thus, RFLPs can be used to pyramid resistance genes, and thereby rapidly identify recessive resistance alleles, or to breed for quarantined pathogens. Moreover, RFLPs can speed backcross breeding for disease-resistance genes by selecting for desirable recombination events near resistance genes, as well as recovering recurrent parent genotype on unlinked chromosomes.

Experiments are now under way to develop a physical map (in which distances between RFLPs are calculated in number of nucleotide base pairs, rather than recombinational units) by pulsed field gel electrophoresis of RFLPs near *Tm-1* and *Tm-2*. A preliminary study (Ganal et al. 1989) has demonstrated that although several RFLPs near *Tm-2* map within a distance of only 1.2 centimorgans of one another, they span a region over three million base pairs in length. These results indicate that rates of recombination are suppressed by a factor of seven near *Tm-2* compared to the tomato genome as a whole. Likewise, many RFLPs linked to *Tm-1* map to a small region of the genome that is known by cytogenetic analysis to be physically very large (Ganal et al. 1988). Therefore, even the RFLPs tightly linked to *Tm-1* or *Tm-2* (in terms of genetic distance) are still very far away in terms of nucleotide base pairs. Map-based cloning of the these genes may be more difficult than originally anticipated.

Conclusion and Summary

Disease-resistance genes have long been recognized solely on the basis of their resistance phenotype. In the past, cloning these genes was not feasible because of the lack of information on their corresponding gene products. Without cloned copies of disease-resistance genes, detailed analysis of their regulation, site of action, or activity in novel genetic backgrounds has been impossible. The development of a new cloning strategy, based on map position rather than prior biochemical information, now makes molecular characterization of resistance genes from tomatoes and other cultivated plants feasible.

The two main obstacles to success in map-based cloning of disease-resistance genes in plants are (1) difficulty in identifying RFLPs that are very tightly linked to target genes, and (2) difficulty in cloning over long distances near tightly linked RFLPs. The use of multiple probing of NILs described here and elsewhere (Young et al. 1988) offers an effective and general approach to rapid identification of tightly linked RFLPs. New techniques, such as pulsed field electrophoresis and cloning into yeast artificial chromosome vectors, greatly simplify long-range cloning.

Another way to facilitate map-based cloning is the use of model plant systems with small and simple genomes, thereby reducing the size and complexity of the chromosomal segment around RFLPs to be cloned and analyzed. In this respect, the tomato is useful because of its relatively small genome. Because of the desirability of carrying out map-based cloning in plants with even smaller and simpler genomes, however, this strategy is now being applied in other model systems. Unfortunately, *Arabidopsis*, the plant species with the smallest genome among higher plants, lacks well-characterized disease-resistance genes. The model legume species, *Vigna*, however, has both a small and simple genome, as well as many extensively characterized disease interactions and resistance genes. One of us (N. Young) is now targeting resistance genes in this organism with tightly linked RFLPs as a foundation for future map-based cloning. These experiments to develop map-based cloning in tomato and other suitable plant systems will soon provide insights into the molecular basis of disease resistance, while creating a foundation for gene cloning in more complex plant systems.

Literature Cited

Burke, D. T., G. F. Carle, and M. V. Olson. 1987. Cloning of large segments of exogenous DNA into yeast by means of artificial chromosome vectors. Science 236:806–812.

Carle, G. F., M. Frank, and M. V. Olson. 1986. Electrophoretic separations of large DNA molecules by periodic inversion of the electric field. Science 232:65–68.

Chu, G., D. Volrath, and R. W. Davis. 1986. Separation of large DNA molecules by contour-clamped homogeneous electric fields. Science 234:1582–1585.

Dunigan, D., D. Golemboski, and M. Zaitlin. 1987. Analysis of the *N* gene of *Nicotiana*. In Plant Resistance to Viruses. Ciba Found. Symp. 133:120–135.

Fraser, R. S. S., and S. A. R. Loughlin. 1980. Resistance to tobacco mosaic virus in tomato: Effects of the *Tm-1* gene on virus multiplication. J. Gen. Virol. 48:87–96.

Ganal, M. W., N. L. V. Lapitan, and S. D. Tanksley. 1988. A molecular and cytogenetic survey of major repeated DNA sequences in tomato (*Lycopersicon esculentum*). Mol. Gen. Genet. 213:262–268.

Ganal, M. W., and S. D. Tanksley. 1989. Analysis of tomato DNA by pulsed field gel electrophoresis. Plant Mol. Biol. Rep. 7:17–28.

Ganal, M. W., N. D. Young, and S. D. Tanksley. 1989. Pulsed field gel electrophoresis and physical mapping of large DNA fragments in the *Tm-2a* region of chromosome 9 in tomato. Mol. Gen. Genet. 215:395–400.

McCormick, S., J. Niedermey, J. Fry, A. Barnason, R. Horsch, and R. Fraley. 1986. Leaf disc transformation of cultivated tomato (*L. esculentum*) using *Agrobacterium tumefaciens*. Plant Cell Rep. 5:81–84.

Miller, J. C., and S. D. Tanksley. 1990. Effects of different restriction enzymes, probe source, and probe length on detecting restriction fragment length polymorphism in tomato. Theor. Appl. Genet. 80:385–389.

Motoyoshi, F., and N. Oshima. 1977. Expression of genetically controlled resistance to tobacco mosaic virus infection in isolated tomato leaf protoplasts. J. Gen. Virol. 34:499–506.

Orkin, S. H. 1986. Reverse genetics and human disease. Cell 47:845–850.

Pelham, H. 1966. Resistance in tomato to tobacco mosaic virus. Euphytica 15:258–267.

Sambrook, J., E. F. Fritsch, and T. Maniatis. 1989. *Molecular Cloning. A Laboratory Manual.* Cold Spring Harbor, NY: Cold Spring Harbor Laboratory Press. 2nd Ed.

Steinmetz, M., K. Minard, S. Harvath, J. McNichols, J. Srenlinger, C. Wake, E. Long, B. Mach, and L. Hood. 1981. A molecular map of the immune response region of the major histocompatibility complex of the mouse. Nature 300:35–42.

Tanksley, S. D., N. D. Young, A. P. Paterson, and M. W. Bonierbale. 1989. RFLP mapping in plant breeding: New tools for an old science. Bio/Technology 7:257–263.

Vallejos, C. E., S. D. Tanksley, and R. Bernatzky. 1986. Localization in the tomato genome of DNA restriction fragments containing sequences homologous to the rRNA (45S), the major chlorophyll a/b binding polypeptide and the ribulose bisphosphate carboxylase genes. Genetics 112:93–105.

Young, N. D., J. C. Miller, and S. D. Tanksley. 1987. Rapid chromosomal assignment of multiple genomic clones in tomato using primary trisomics. Nucl. Acid Res. 15:9339–9348.

Young, N. D., D. Zamir, M. W. Ganal, and S. D. Tanksley. 1988. Use of isogenic lines and simultaneous probing to identify DNA markers tightly linked to the *Tm-2a* gene in tomato. Genetics 120:579–585.

Young, N. D., and S. D. Tanksley. 1989. RFLP analysis of the size of chromosomal segments retained around the *Tm 2* locus of tomato during backcross breeding. Theor. Appl. Genet. 77:353–359.

CHAPTER 6

Genetics of Resistance to Viral Diseases of Bean

ROSARIO PROVVIDENTI

Introduction

Bean (*Phaseolus vulgaris*) and its close relatives, lima bean (*P. lunatus*), tepary (*P. acutifolius*), and runner bean (*P. coccineus*), are of American origin. Archeological remains indicate that they have been cultivated for thousands of years. A few weeks after landing in the New World (1492), Columbus admired well-kept bean fields in Cuba. These "faxones" or "fabas" appeared to be very different from those grown in Spain and in other Mediterranean countries. Hence, the Age of Discovery opened new frontiers to the common bean and its relatives. A century later, the cultivation of beans had spread rapidly (considering the slow methods of transportation prevailing at that time) through Europe. The botanists Clusius (1576) and Bauhim (1616) in their books *Plantarum Historia* and *Historia Plantarum* described a number of bean varieties under the name of *Phaseoli peregrini* (foreign beans). In 1629, Parkinson noted that the American beans were "rich man's food," too expensive for the common people (Hedrick et al. 1931).

In 1753, Linneaus classified the common bean as *P. vulgaris*, the lima bean as *P. lunatus*, and the runner bean as *P. coccineus*. Later, however, other botanists, impressed by the great genetic diversity

Dr. Provvidenti is in the Department of Plant Pathology, Cornell University, New York State Agricultural Experiment Station, Geneva, New York 14456.

among bean varieties, proceeded to describe a number of new botanical species. The new names tended to be more descriptive regarding seed color, seed shape, and place of origin. Hence the new species: *Phaseolus mexicanus*, *P. oblongus*, *P. pictus*, *P. angulosus*, *P. ovalispermus*, *P. subglobosus*, *P. nigricans*, *P. saponaceus*, and many others. These botanists failed to take into consideration the genetic diversity within a species; consequently, after 248 years, we consider valid only the classification given to us by the father of modern botany (Hedrick et al. 1931).

Biochemical and genetic evidence indicates that there are two centers of domestication for *P. vulgaris*: the MesoAmerican and the Andean. Landraces from these two centers differ significantly in plant type, growth habit, seed size, disease resistance, phaseolins, lectins, and mitochondrial DNA (mtDNA). Genetic incompatibility is one major feature affecting members of these two groups. For example, two distinct genes have been identified and are known as *DL1* (*Dwarf Lethal 1*), present in the small-seeded beans belonging to the MesoAmerican center, and *DL2* (*Dwarf Lethal 2*), in the large-seeded land races from the Andean region (e.g. kidney, cranberry, snap beans, and others). Plants of the F_1 generation that derived from parents possessing these two genes are mostly chlorotic, stunted, and frequently die prematurely. Other difficulties and limitations have been experienced during breeding programs involving members of these two landrace groups (Provvidenti and Schroeder 1969; Brick et al. 1990).

Field (dry) and garden (snap) beans presently constitute one of the most important food crops of the world, and they are grown in almost every country. The world production of dry beans is about 15 million tons; one fourth of that is grown in Latin America, another fourth in the Far East, and the rest in many other countries. India, China, Brazil, Mexico, and the USA are the leading producers. In less developed countries, dry beans rank high as an economical source of nourishing food because they are high in protein, calcium, iron, and vitamin B1 (Schultz and Dean 1947; Brick et al. 1990).

Through the years, thousands of varieties have been developed by selection or through breeding. These varieties may differ in maturity, adaptation to particular climatic and soil conditions, size, shape, and color of the seeds. At times, a relatively minor physiological or morphological trait may have major economic importance. The increasing requirements of the food processing industry are also reflected in the new varieties. Typically, modern varieties are dwarf or bush with pods set in the central part of the plant situated well up from the soil, and uniform in maturity, all necessary elements for mechanical harvesting. This accumulation of desirable characters has also eliminated many other characters associated with the wild types, which would confer field tolerance to diseases and insects. Consequently, bean crops, as other highly developed vegetables, are more susceptible to fungi, bacteria, and viruses than the old landraces of several decades ago (Schwartz and Pastor-Corrales 1989; Brick et al. 1990).

This chapter focuses on diseases caused by viral agents, and on sources of heritable resistance, genetics, and breeding.

Table 1. *Viruses of* Phaseolus vulgaris.

VECTOR & VIRUS	GROUP	SYMPTOMS	RESISTANCE GENE
Aphids			
Alfalfa mosaic	AMV group	Mosaic, yellow dots	*Amv, Amv-2*
Bean common mosaic	Potyvirus	Green mosaic, cupping, stunting	*I, bc-1, bc-1*^{2} *bc-2, bc-2*^{2} *bc-3, bc-u*
Bean leaf roll	Luteovirus	Mosaic, distortion	—
Bean yellow mosaic	Potyvirus	Yellow mosaic	*By-2*
Broad bean wilt	Fabavirus	Yellow mosaic, distortion	*Bbw*
Blackeye cowpea mosaic	Potyvirus	Mosaic, necrosis, wilting	*Bcm*
Clover yellow vein	Potyvirus	Yellow mosaic, necrosis, wilting	*cyv*
Cowpea aphid-borne mosaic	Potyvirus	Mosaic, necrosis, wilting	*Cam*
Cucumber mosaic	Cucumovirus	Green mosaic, blisters	°
Passionfruit woodiness	Potyvirus	Mosaic, blisters, distortion	*Pwv*

° = Sources of resistance are available.

Table 1. Continued.

VECTOR & VIRUS	GROUP	SYMPTOMS	RESISTANCE GENE
Pea mosaic	Potyvirus	Yellow mosaic	*By*
Peanut mottle	Potyvirus	Necrosis, wilting	*Pmv*
Soybean mosaic	Potyvirus	Green mosaic, stunting	*Smv*
Watermelon mosaic	Potyvirus	Yellow mosaic	*Wmv, Hsw*
Beetles			
Bean curly dwarf mosaic	Ungrouped	Mosaic, rugosity, stunting	*
Bean mild mosaic	Ungrouped	Green mosaic	*
Bean pod mottle	Comovirus	Mosaic, rugosity	*Bpm*
Bean rugose mosaic	Comovirus	Mosaic, severe rugosity	*Mrf*, Mrf^2
Bean southern mosaic	Sobemovirus	Green mosaic, rugosity	*Bsm*
Blackgram mottle	Carmovirus	Mottle, distortion	*
Cowpea chlorotic mottle	Bromovirus	Yellow spots	—
Fungi			
Tobacco necrosis virus	Necrovirus	Necrosis, stunting	—
Leafhoppers			
Beet curly top	Geminivirus	Curling, yellowing, stunting	*Ctv, ctv-2*
Tobacco yellow dwarf virus	Ungrouped	Yellowing, wilting	*
Nematodes			
Tobacco ringspot	Nepovirus	Mosaic, necrosis, stunting	*trv*
Tomato ringspot	Nepovirus	Mosaic, necrosis, stunting	*
Thrips			
Tobacco streak virus	Ilarvirus	Red nodes, red spots	*

* = Sources of resistance are available.

Table 1. Continued.

VECTOR & VIRUS	GROUP	SYMPTOMS	RESISTANCE GENE
Whiteflies			
Bean golden mosaic	Geminivirus	Golden mosaic, stunting	*
Bean dwarf mosaic	Ungrouped	Yellow mosaic, stunting	*
Euphorbia mosaic	Geminivirus	Necrotic lesions, distortion	*
Rhynchosia mosaic	Unknown	Yellowing, stunting	*
Unknown			
Tobacco mosaic virus	Tobamovirus	Necrotic local lesions	*Tm*

* = Sources of resistance are available.

Viruses Transmitted by Aphids

A number of aphid-transmitted viruses infect bean (*Phaseolus vulgaris*) and related species. This section discusses those of great economic importance as well as those for which the literature is limited regarding their occurrence and damage to bean crops. Although some of these diseases appear to be of secondary economic importance, their potential seriousness should not be disregarded. This group includes alfalfa mosaic (AMV), bean common mosaic (BCMV), bean leaf roll (BLRV), bean yellow mosaic (BYMV), broad bean wilt (BBWV), blackeye cowpea mosaic (BlCMV), cowpea aphid-borne mosaic (CAMV), cucumber mosaic (CMV), passion-fruit woodiness (PWV), pea mosaic (PMV), peanut mottle (PMV), soybean mosaic (SMV), and watermelon mosaic (WMV). All these viruses are spread by a number of aphid species (*Aphis craccivora*, *A. fabae*, *A. gossypii*, *Aphis glycine*, *A. medicagins*, *A. rumicis*, *Acyrthosiphon pisum*, *Aulacorthum solani*, *Hyalopterus atriplicis*, *Macrosiphum euphorbiae*, *Myzus persicae*, *Rhopalosiphum pseudo-*

brassicae, and others). Most of the viruses mentioned above are transmitted in the nonpersistent manner and, in general, virus acquisition and transmission occurs within a minute. A few of these viruses are seedborne in *P. vulgaris* and other *Phaseolus* species (Pierce 1935; Bos 1970; Bos 1972; Taylor and Stubs 1972; Taylor and Greber 1973; Alconero and Meiners 1974; Bock and Conti 1974; Hollings and Stone 1974; Bock and Kuhn 1975; Francki et al. 1979; Jasper and Bos 1980; Purcifull et al. 1984; Purcifull and Gonsalves 1985; Morales and Bos 1988). Methods of control usually include the eradication of overwintering hosts of the virus, and applications of insecticides to eliminate aphid vectors. In some cases, insecticides can reduce the secondary spread of the virus from the primary foci of infection, but generally they do not prevent or eliminate viral disease. Consequently, the use of resistant varieties is undeniably the most efficient and economical method of viral disease control.

Alfalfa Mosaic Virus

Alfalfa mosaic virus is a virus with bacilliform particles of different lengths, 30 to 60 nm, and a diameter of 15 nm in which four single strands of RNA of messenger polarity are separately packaged. The three largest RNAs comprise the genome; the fourth is a subgenomic message for the coat protein. The three genome RNAs and either the fourth RNA or the coat protein are needed for infectivity. The virus is readily sap-transmissible, seed-transmitted in some hosts (e.g. alfalfa and pepper), and is transmitted in the nonpersistent manner by at least 14 aphid species to a very wide range of host plants. It is of world-wide distribution, and many strains have been reported or characterized. It occurs naturally in many wild and cultivated species, often without symptoms. Infection may be latent or masked, and recovery often occurs. Symptoms depend greatly on the virus strain, host variety, stage of growth, and environmental conditions. In many species, these include mottle or mosaic, which can be very bright yellow (calico). Local and systemic necrosis may also occur. In *P. vulgaris*, many strains incite necrotic local lesions that sometimes coalesce, appearing 3 to 5 days after inoculation. Other strains cause chlorotic local lesions followed by a systemic mottle, vein necrosis, and leaf distortion (Jasper and Bos 1980). Wade and Zaumeyer (1940) found that resistance was conditioned by two independently inherited genes (possibly duplicate entities), which conferred a very high level of resistance to the same strain of the virus. To these genes, the symbols *Amv* and *Amv–2* were later assigned (Provvidenti 1987).

Bean Common Mosaic Virus

Bean common mosaic virus contains a monopartite RNA genome encapsidated in flexuous particles about 750 nm long and 15 nm in diameter typical of the potyvirus group. It was recognized in 1917 as the causal agent of a major bean disease by Stewart and Reddick (1917). The virus is distributed worldwide, due mainly to its transmission in seed of common bean (*Phaseolus vulgaris*) and other species (*P. aborigineus*, *P. coccineus*, *P. acutifolius*, *P. angustifolius*, *P. polyanthus*, *Macroptilium lathyroides*, and *Vigna radiata*) (Hoch and Provvidenti 1978; Kline et al. 1988). BCMV is readily transmitted by several species of aphids and by mechanical means (Reddick and Stewart 1919; Morales and Bos 1988). Because BCMV infections reduce the quality and yield or even cause total loss of the crop, it can be extremely damaging. Most of the isolates of the virus cause plant stunting and a mosaic consisting of a foliar mottle, green veinbanding, curling, blisters, and distortion. In certain resistant plants, some strains of the virus can cause a disease known as "black root," consisting of vascular necrosis that induces wilting and premature death of infected plants. In general, the intensity and severity of symptoms depend upon the variety, virus strain, age of plants, and environmental conditions (Alconero and Meiners 1974; Morales and Bos 1988).

To prevent virus spread by seeds, the USA and Canada require the use of certified seed that guarantees infection of 0% for foundation stocks and less than 1% for commercial lots. No international agreement is available, and BCMV is still commonly found in varieties of foreign introduction. The rate of seed transmission may vary from 3% to 83%. The virus has been detected in pollen and embryos of dormant and germinating bean seeds, and no chemical or other treatment can inactivate virus particles within the seed. In some cases, insecticides may reduce a secondary spread of the virus from primary sites of infection by controlling the vectors. BCMV can be effectively eliminated, however, by the use of resistant varieties.

Resistance to BCMV was discovered in the early 1930s when researchers selected resistant plants from some of the commercial varieties. One type of resistance found in 'Robust' and 'Great Northern 1' was inherited recessively (Pierce 1935). A second type of resistance was located in the variety 'Corbett Refugee' and inherited dominantly (Ali 1950). Ali was the first to propose a genetic explanation of the inheritance of these two types of resistance. He used the symbol *aa* for the recessive gene and the symbol *AAII* for the dominant factor. Ali explained that the factor *A* was required for infection, and the factor *II* would inhibit symptom expression (Ali 1950). The scheme of Ali was amended by Peterson (1958) in Germany, who introduced additional gene symbols and indicated that susceptible, resistant, and hypersensitive reactions were related to the strain of the virus. He and other researchers determined that resistance was strain-specific, and that it was important to classify the different strains of the virus. In 1978, Drijfhout in Holland, in cooperation with Silbernagel and Burke in the USA, proposed sets of differential bean lines that can be used to differentiate these strains (Drijfhout et al. 1978).

Strains of BCMV were classified into ten pathogenicity groups indicated by Roman numerals: I, II, III, IVa, IVb, Va, Vb, VIa, VIb, and VII (Drijfhout 1978). These strains or pathotypes were also grouped into three main categories:

1. Mosaic-inducing strains, which cause only systemic mosaic.
2. Temperature-dependent necrosis-inducing strains, causing systemic necrosis only at high temperatures (above 30°C.) in plants possessing the *II* gene alone or in combination with some recessive genes.
3. Temperature-independent necrosis-inducing strains, which induce systemic necrosis at any temperature in plants possessing the *II* gene alone or in combination with some recessive genes.

Plants with the *II* gene will develop necrosis at any temperature with any strain of the virus if grafted to infected plant material. Strains causing only mosaic, and temperature-dependent necrosis-inducing strains, are of American origin and probably coevolved with

Phaseolus species at their center of origin. A recent study conducted by Kline et al. (1988) showed that only strains of Serotype B (mosaic-inducing strains) were found in seeds of the *Phaseolus* germplasm collection at the Western Regional Plant Introduction Station, Pullman, Washington. The temperature-independent necrosis-inducing strains identified and characterized by researchers in the Netherlands (NL) appear to be of African origin. These strains, known as NL-3, NL-5, and NL-8, are very destructive in plants possessing the *II* gene alone or in association with some recessive genes (Hubbeling 1972; Drijfhout and Bos 1977). In susceptible varieties, however, they incite foliar mosaic similar to that caused by other strains of BCMV, although they differ serologically (Serotype A) (Wang et al. 1982).

These African strains were introduced in the USA, via Europe, most likely through infected seeds. NL-3 and NL-8 have been found in bean fields of Idaho, Washington, Montana, New York, and Ontario and are presently causing a great deal of concern to seedsmen and growers (Provvidenti et al. 1984).

Drijfhout (1978) proposed replacing the symbol *a* given by Ali with *bc* to abbreviate bean common. From the results of many diallel crosses of differential varieties possessing recessive factors, he reached the following conclusions:

1. There are six recessive genes.
2. One, *bc-u*, is strain-unspecific and complementary to a series of strain specific genes.
3. The five strain-specific genes are at three loci: *bc-1*, $bc\text{-}1^2$, *bc-2*, $bc\text{-}2^2$, and *bc-3*.
4. The sets of genes *bc-1*/ $bc\text{-}1^2$ and *bc-2*/$bc\text{-}2^2$ are either allelic or very tightly linked.
5. Resistance occurs only when *bc-u* is present with at least one of the strain-specific genes.

The following differential varieties used for strain identification were found to possess one or more strain-specific genes and the unspecific genes:

'Imuna', 'Puregold Wax', 'Redland Greenleaf C'	*bc-u*	*bc-1*		
'Redland Greenleaf B', 'Great Northern UI 123'	*bc-u*	*bc-1*^{2}		
'Sanilac', 'Michelite 62', 'Red Mexican UI 34'	*bc-u*		*bc-2*	
'Pinto UI 114'	*bc-u*	*bc-1*	*bc-2*	
'Monroe', 'Red Mexican UI 35', 'Great Northern UI 31'	*bc-u*	*bc-1*^{2}	*bc-2*^{2}	
'IVT 7214'	*bc-u*		*bc-2*	*bc-3*

It is evident that a number of other combinations are possible through breeding, which could be used to control yet undiscovered strains of BCMV:

bc-u	*bc-1*	*bc-2*	*bc-3*
bc-u	*bc-1*	*bc-2*^{2}	*bc-3*
bc-u	*bc-1*^{2}	*bc-2*	*bc-3*
bc-u	*bc-1*^{2}	*bc-2*^{2}	*bc-3*
bc-u		*bc-2*^{2}	*bc-3*
bc-u	*bc-1*^{2}		*bc-3*
bc-u	*bc-1*		*bc-3*
bc-u	*bc-1*	*bc-2*^{2}	
bc-u			*bc-3*
bc-u		*bc-2*^{2}	

Drijfhout crossed a number of varieties possessing the *II* gene alone, or in combination with some of the recessive genes, and elucidated the genetic configuration of the following differential lines:

'Widusa', 'Black Turtle Soup'	*I*		
'Jubila', 'Topcrop'	*I*	*bc-1*	
'Amanda'	*I*	*bc-1*^{2}	
'IVT 7233'	*I*	*bc-1*^{2}	*bc-2*^{2}

The varieties 'Black Turtle Soup' and 'Widusa' were found to possess only the *II* gene, whereas the others included one or more of the recessive factors. It appears that Drijfhout was not able to distinguish plants with or without *bc-u*. He has hypothesized that this gene in the homozygous condition and in association with strain-specific genes confers a higher level of resistance than those without it. Again, through breeding it is possible to generate other combinations, which could efficiently control undiscovered strains or new mutations of BCMV.

I		*bc-2*	
I		*bc-2*^{2}	
I		*bc-3*	
I	*bc-1*	*bc-2*	
I	*bc-1*	*bc-2*^{2}	
I	*bc-1*		*bc-3*
I	*bc-1*^{2}	*bc-2*	
I	*bc-1*^{2}		*bc-3*
I		*bc-2*	*bc-3*
I		*bc-2*^{2}	*bc-3*
I	*bc-1*	*bc-2*	*bc-3*
I	*bc-1*	*bc-2*^{2}	*bc-3*
I	*bc-1*^{2}	*bc-2*	*bc-3*
I	*bc-1*^{2}	*bc-2*^{2}	*bc-3*

The practical implications of breeding for BCMV resistance by using the *II* gene and other recessive characters include the following:

1. The *II* gene is rather versatile and provides a wide range of resistance; however, since it conditions hypersensitivity, this resistance can be broken by temperature-sensitive or insensitive strains or by grafting.
2. The *II* gene alone can control only certain strains of BCMV, thus for wide protection some of the recessive strain-specific genes must be present.
3. Some strains are seedborne in plants possessing recessive genes, but none is seed- transmitted in varieties with *II*, because necrotic plants die prematurely.
4. Aphids can introduce the virus in an *II* crop only from nearby susceptible crops. Virus transmission by aphids from plants with

systemic necrosis has never been observed, thus further viral spread within an *II* crop does not occur.

5. The exclusive cropping of *II* would completely prevent the survival and spread of BCMV, since beans and other related species are practically the only natural sources of the virus. In some areas of Africa where NL-3, NL-5, and NL-8 originated, these strains probably are endemic in other hosts and are a constant threat to the bean crops.
6. Two lines developed by Drijfhout of the Institute for Horticultural Plant Breeding (IVT) of Holland, 'IVT 7214' homozygous for (*bc-u*, *bc-2*, *bc-3*), and 'IVT 7233' (*I* $bc\text{-}1^2$ $bc\text{-}2^2$), are resistant to all known strains of BCMV, but apparently they are not available for distribution. A multiresistant line with the same genetic configuration of 'IVT 7233' can be easily obtained, however, by crossing the variety 'Great Northern 31', 'Monroe', or 'Red Mexican UI 35' ($bc\text{-}1^2$, $bc\text{-}2^2$) with 'Black Turtle Soup', 'Alliance', or 'Blazer', which possess only the *II* gene.
7. Tests conducted in Africa under natural conditions and further experimental screening at Centro Internacional por Agricultura Tropical (CIAT) revealed 98 BCMV-resistant lines. Of particular value are lines MCM 251, MCM 2201-1, MCR 2501-15, and VCA 81018, which remained symptomless when inoculated with NL-3 and NL-5.
8. Through breeding, a number of theoretical combinations are possible by combining the *II* genes with recessive strain-specific genes. These can be of great value in controlling new pathotypes or mutants of the virus that will be recognized in the future.

Bean Leaf Roll Virus

Bean leaf roll virus is a luteovirus with isometric virus particles about 27 nm long containing a single RNA component. It is transmitted by several species of aphids in a persistent manner. There is a minimum acquisition access period of 2 hours, a minimum latent period of 100 or more hours, and a minimum inoculation access period of 15 minutes. The virus does not multiply in the vector, but it can be retained for 15 or more days. As other viruses of this group, BLRV is neither mechanically nor seed transmitted. Serologically it is related to soybean dwarf virus, beet western yellows virus, subterranean clover red leaf virus, potato leaf roll virus, and carrot redleaf virus. This virus infects mostly leguminous species and can be a serious problem in pea and fava bean. In bean, it appears to be of minor importance, and the symptoms incited by this virus can be easily confused with those incited by other viral agents. These include plant stunting, yellowing, and rolling and rigidity of affected leaves (Ashby 1984).

Very little is known regarding resistance or tolerance to this virus. Application of insecticides could delay infection and somewhat protect the bean crops, if sprayed at the right time. Other sanitation measures could also be useful.

Bean Yellow Mosaic Virus

Bean yellow mosaic virus is a member of the potyvirus group based on its particle size, ability to induce typical cytoplasmic inclusions in the host cells, immunological relationships with other members of its group, and aphid transmission. Virus particles are long flexuous rods averaging 750 nm long and 15 nm wide and containing a single strand of RNA. Serologically, BYMV is related to a number of other potyviruses, such as bean common mosaic, clover yellow vein, soybean mosaic, watermelon mosaic virus, wisteria vein mosaic, and others. BYMV can infect most of the species of the Leguminosae family, including herbaceous (*Cajanus*, *Canavalia*, *Cassia*, *Cicer*, *Crotolaria*, *Dolichos*, *Glycine*, *Hedysarum*, *Lathyrus*, *Lens*, *Lupinus*, *Medicago*, *Melilotus*, *Phaseolus*, *Pisum*, *Trifolium*, *Trigonella*, *Vicia*, *Vigna*, etc.), woody plants (*Robinia* and *Cladrastis*), and non-legume species (*Alpinia*, *Chenopodium*, *Gladiolus*, *Freesia*, *Babiana*, *Sparaxis*, and *Tritonia*). A number of variants, strains, and pathotypes have been characterized through the years; however, some of the most notable strains are now considered to be distinct viral entities. The virus is also easily mechanically transmissible, thus facilitating breeding for resistance and other studies. No seed transmission has been demonstrated for BYMV in *Phaseolus vulgaris* or any of the related *Phaseolus* spp., but it is known to occur in *Vicia faba*, *Trifolium pratense*, *Lupinus albus*, and *L. luteus*. BYMV is known to be of worldwide distribution, occurring in most of the countries where beans and other legumes are cultivated. BYMV has been reported to be the agent of devastating epidemics, causing considerable losses in yield and quality of the bean crop, and also to infect entire fields with only minor damage. Several factors usually determine the severity of this and other viral diseases, e.g. varieties, plant age, vectors, strains of the virus, and environmental factors (Pierce 1934; Bos 1970).

Typical foliar mosaic consists of contrasting dark and yellowish green areas, often accompanied by bright yellow spots, which intensify as the plant ages. Some isolates incite only a mild and diffuse chlorotic mottle and limited plant stunting, whereas others cause a prominently coarse mosaic, rugosity, malformation, and severe stunting. Some varieties may respond also with necrotic spots, veinal and apical necrosis, wilting, and premature death. Generally, a late infection causes less prominent foliar symptoms, and if pod development coincides with infection, the pods eventually will exhibit a light green mottle and slight malformation. Pod symptoms are often of diagnostic value in distinguishing BYMV infection from that caused by clover yellow vein virus (CYVV). The latter virus, formerly known as the severe strain of bean yellow mosaic virus (BYMV-S), causes very prominent pod distortion (Provvidenti and Schroeder 1973). BYMV-infected plants are usually stunted and bushy, due to reduction of internode length and proliferation of lateral branches. Plant maturity is delayed and the quality and quantity of seed production are frequently affected (Pierce 1934; Bock and Kuhn 1975).

Resistance can significantly reduce the losses caused by this virus, particularly in dry-bean varieties, which, during their long vegetative cycle, are exposed for months to viral infections. The major genetic factor presently available derived from the interspecific cross *P. coccineus* × *P. vulgaris*. Dickson and Natti (1968) determined that a single dominant gene, *By-2*, is able to confer resistance to most of the isolates and strains of the virus. A second source of strain-specific resistance was found in 'Great Northern 31'. Tatchell et al. (1985) reported that resistance in this variety is conferred by three complementary recessive genes. Some bean varieties are able to escape or delay BYMV infection without possessing a specific factor for BYMV resistance; however, their performance varies with the locality in which they are grown. This inconsistency is probably due to the occurrence of different aphid species or specific biotypes, or vector preference. Resistance conferred by specific genetic factors offers a reliable and durable form of control for many viral diseases. Unquestionably, resistant varieties with the *By-2* gene can significantly reduce the incidence of losses to this virus.

Broad Bean Wilt Virus

Broad bean wilt virus has isometric particles about 25 nm long containing a single-stranded RNA. Although it shares some similarities with cowpea mosaic virus, it differs in other properties, hence it is classified in the fabavirus group. In nature it is spread by several aphid species, but it also can be easily transmitted mechanically. There is no evidence of seed transmission. There are several strains of this virus, which can be grouped in two serotypes. BBWV can infect a large number of plant species, including several ornamentals. It has been found to cause severe epidemics in spinach, lettuce, pea, broad bean (fava bean), and other legumes, including bean. There are very few reports in the literature regarding the natural occurrence of BBWV in beans, perhaps because the symptoms caused by this virus on naturally infected plants resemble those caused by other viruses. These symptoms include chlorotic concentric ringspot, severe foliar distortion, and plant stunting. Some of the kidney type and yellow wax type beans appear particularly susceptible, but there are varieties of these two types of bean that are resistant (Taylor and Stubs 1972; Uyemoto and Provvidenti 1974).

This virus can be effectively controlled by using resistant varieties. A large number of bean varieties of different types were found to be resistant. In crosses between the resistant variety 'Sanilac' and the susceptible 'Red Kidney', resistance was determined to be controlled by the single dominant factor *Bbw* (Provvidenti 1988a).

Blackeye Cowpea Mosaic Virus

Blackeye cowpea mosaic virus (BlCMV) is a potyvirus with filamentous particles about 750 nm long containing a single-stranded RNA. Serologically it is closely related to some strains of bean common mosaic virus and peanut stripe virus, and is distantly related to cowpea aphid-borne mosaic virus (CAMV). It is aphid-, mechanically, and seed-transmitted. Seed transmission in cowpea can be as high as 40%, thus this route is the probable major source of this virus. BlCMV can also infect a number of other leguminous species, including the common bean. In this host, symptoms consist of veinal chlorosis, mottle, and stunting. Older leaves tend to develop an intense chlorosis, wilt, and abscission. Plants usually die prematurely without producing any seed, hence no information is available regarding the seedborne nature of this virus in bean. Nor is information available regarding the natural occurrence of BlCMV in bean. Nevertheless, if a susceptible variety is grown in the vicinity of an infected cowpea field, bean plants can be devastated (Purcifull and Gonsalves 1985).

A large number of bean varieties were found to be resistant to this virus. In the variety 'Black Turtle 1', this resistance was found to be conferred by the single dominant gene *Bcm* (Provvidenti et al. 1983). Varieties resistant to BlCMV are resistant also to CAMV, thus similarities in symptomatology, nature of resistance, and mode of inheritance, and an apparent close linkage between the resistance factor for these two viruses, suggest a common gene or two closely linked genes. Genetic studies suggest a relationship between these factors and the *I* gene (Kyle 1988). When no resistant variety is available, it is recommended to avoid growing bean near cowpea. The use of insecticides could delay the rapid spread of these viruses.

Clover Yellow Vein Virus

As all the other potyviruses, clover yellow vein virus (CYVV) possesses filamentous virus particles about 760 nm long and 12 nm wide, containing a single strand of RNA. Serologically, it is related to BYMV, lettuce mosaic, soybean mosaic, turnip mosaic, and a few others. CYVV incites the development of cytoplasmic and nuclear inclusions in cells of infected hosts. The host range of CYVV is broader than that of BYMV, since it can infect most of the legumes (*Cajanus*, *Canavalia*, *Cassia*, *Cicer*, *Crotolaria*, *Dolichos*, *Glycine*, *Hedysarum*, *Lathyrus*, *Lens*, *Lupinus*, *Medicago*, *Melilotus*, *Phaseolus*, *Pisum*, *Trifolium*, *Trigonella*, *Vicia*, *Vigna*, etc.) as well as a large number of species of certain genera (*Antirrhinum*, *Atriplex*, *Chenopodium*, *Coriandrum*, *Cucurbita*, *Gladiolus*, *Gomphrena*, *Nicotiana*, *Nicandra*, *Papaver*, *Petunia*, *Proboscidea*, *Rubus*, *Spinacia*, *Tetragonia*, *Viola*, etc). CYVV was reported first from white clover grown in Surrey, England (Hollings and Stone 1974). Many years before, one of the most important strains of this virus had been categorized by Grogan and Walker (1948) as the pod-distorting strain of bean yellow mosaic virus, and by others as the necrotic or the severe strain of this virus (BYMV-S). Clover yellow vein disease is very destructive to bean crops and is widespread throughout the world. Attempts to demonstrate seed transmission of CYVV in bean, fava, pea, and other legume species have been unsuccessful (Hollings and Stone 1974).

Symptoms caused by CYVV in common bean are variable, depending upon the bean variety, strain of the virus, time of infection, and environmental conditions. The most prevalent strain causes prominent yellow mosaic, malformation, and pronounced plant stunting. A number of varieties also respond with apical necrosis, premature defoliation, wilting, and plant death. These plants are frequently attacked by soilborne pathogens and develop severe root-rot, which can be easily mistaken as the primary agent of the disease. Although the yellow foliar mosaic resembles that incited by strains of BYMV, pods are severely distorted and mottled, rendering them worthless for fresh market, canning, or freezing (Grogan and Walker 1948; Hollings and Stone 1974).

Resistance in bean to CYVV was first reported by Grogan and Walker (1948) to occur in Great Northern varieties. In 1973, Provvidenti and Schroeder established that resistance in 'Great Northern 1140' is governed by the single recessive gene *by-3*, to which a new symbol, *cyv*, was later assigned (Provvidenti 1987). In 1983, Tu (1983) found that resistance in the navy bean variety 'Clipper', was also monogenically recessive. Recently, Tatchell et al. (1985) demonstrated that in 'Great Northern 31', resistance is conferred by two recessive complementary genes. Thus, it is evident that resistance to CYVV in *P. vulgaris* is governed by different genetic factors, but it is not strain-specific. Sources of resistance to CYVV are also available in some accessions of *P. coccineus*.

Cowpea Aphid-Borne Mosaic Virus

Cowpea aphid-borne mosaic virus (CAMV) is a potyvirus with flexuous rods about 750 nm long containing a single-stranded RNA. It is transmitted by several species of aphids and mechanically. It is seed-transmitted in cowpea (*Vigna unguiculata*); consequently it can be found wherever this crop is cultivated. In cowpea, it incites chlorotic mottle, dark green veinbanding with interveinal chlorosis, leaf distortion, and some blistering. This virus can infect many Leguminosae species, including bean. In *P. vulgaris*, symptoms consist initially of veinal chlorosis, which is followed by a diffuse chlorosis, necrosis, wilting, and eventually death. Very little information is available regarding the occurrence and distribution of this virus in bean; however, if susceptible varieties are cultivated near cowpea fields, they may be readily infected, with disastrous consequences (Bock and Conti 1974).

A large number of bean varieties were found to be resistant to this virus. In resistant plants, infection is confined to the inoculated leaves, since the virus fails to move systemically (systemic resistance). In 'Black Turtle 1', resistance was found to be conditioned by the single dominant gene *Cam* that is associated with the *I* gene (Provvidenti et al. 1983; Kyle 1988).

Cucumber Mosaic Virus

Cucumber mosaic virus (CMV) is a cucumovirus with icosahedral particles about 28 nm in diameter with a genome divided among three single-stranded, positive-sense RNAs (RNAs 1, 2, and 3). The subgenomic RNA 4, which codes for the coat protein production, is a duplication of the sequences present in RNA 3. CMV is often associated with replication-dependent small RNAs, called satellites, which usually contain 334–342 nucleotides that have little homology to the genomic RNAs of the virus. Satellites, by definition, depend on CMV for their replication and are encapsidated in its particles. In some crops, satellites can greatly affect symptom development by the CMV genomic RNAs. CMV is transmitted by more than 60 aphid species in the nonpersistent manner with varying degrees of efficiency. The virus is usually acquired by aphid instars within one minute, but their ability to transmit it declines rapidly and is lost within a few hours. CMV is easily transmitted mechanically and, although there is no evidence that it is seed-borne in any of the cucurbits, it has been detected in the seeds of 19 other plant species. CMV is of world-wide distribution and can infect about 800 plant species. There is a particular strain of CMV that systemically infects legumes and is often referred to as the legume strain of CMV (CMV-L). It is able to infect bean, lima bean, pea, cowpea, and other leguminous species systemically. In early stages of plant growth, systemic symptoms consist of a prominent leaf epinasty, followed by a mosaic confined usually to a few leaves, and then complete recovery. Foliar symptoms depend upon the variety, and they include leaf curling, green or chlorotic mottle, blisters, dark green veinbanding, and a zipper-like rugosity along the main veins. Some varieties respond with leaf deformity that can be easily confused with that caused by herbicide injury. Although plants recover from symptoms, the virus continues to replicate in symptomless growth, preventing re-infection. In plants that have reached the blooming stage, symptoms, if present, are confined to the apical leaves, but pods are mostly reduced in size, curved, and mottled. Although entire fields may be infected, economic losses can vary greatly, depending upon the stage of crop growth. If plants are infected when young, they often recover quickly from the initial shock reaction caused by the CMV infection and subsequently produce normally. When infection coincides with development of pods, however, these are mostly misshapen and thus unsuitable for canning or freezing (Whipple and Walker 1941; Bird et al. 1974; Bos and Maat 1974; Provvidenti 1976; Francki et al. 1979; Davis and Hampton 1986).

No sources of resistance have been found in accessions of *Phaseolus vulgaris*, but a number of other species are resistant, including *P. acutifolius*, *P. adenanthus*, *P. anisotrichus*, *P. polyanthus*, *P. trilobatus*, and some accessions of *P. coccineus*. No information is available regarding the mode of resistance in these species (Provvidenti 1976). The most common method of control consists of reducing the presence of aphid vectors in the bean crop.

Passionfruit Woodiness Virus

Passionfruit woodiness virus (PWV) is another potyvirus with rod-shaped virus particles (750 nm) containing a single strand of RNA. Under this name it appears there are a number of related viruses that differ in host range and other properties, but all cause the same symptoms on fruits of *Passiflora edulis* consisting of malformation and a thick, hard pericarp. PWV is spread by a number of aphid species, and can be transmitted mechanically, but there is no evidence of seed transmission. Although PWV occurs in a number of wild and cultivated legumes, very little is known regarding the economic impact of this virus in common bean, in which symptoms are very prominent and destructive. They include plant stunting, leaf reduction accompanied by mosaic, leaf cupping, prominent blisters, and distortion. Pods are also mottled and reduced in size, and seeds are discolored (Taylor and Greber 1973).

Very recent research has shown that a number of varieties respond with local infection without systemic spread (systemic resistance). This resistance is conferred by single dominant genes, which are strain-specific. 'Great Northern 1140' was found to be resistant to a strain from Puerto Rico, but susceptible to another from Australia. Conversely, 'Black Turtle 1' was resistant to both strains (Provvidenti et al. 1992). Sanitation measures and the application of insecticides may limit the spread of this virus, but much depends upon the particular local conditions.

Pea Mosaic Virus

Pea mosaic virus (PMV) is a potyvirus with flexuous, rod-shaped particles containing a single strand of RNA. For several years, it was considered to be a strain of bean yellow mosaic virus, with which it shares some common characteristics. Recent work with RNA/cDNA hybridization and quantitative serology indicate only a distant relationship (Barnett et al. 1987). This virus is spread by a number of aphid species, and is mechanically transmissible. No evidence of seed transmission is available. PMV infects a large number of leguminous species, but only a limited number of bean varieties are infected by this virus, hence it appears to be of minor importance. Symptoms are similar to those caused by BYMV, consisting of a yellow mosaic with some foliar malformation (Pierce 1934; Bos 1970).

It appears that most of the commercial bean varieties are resistant to PMV. In the variety 'California Light Red Kidney', it was determined that this resistance is conferred by a single dominant gene, designated *By* (Schroeder and Provvidenti 1970).

Peanut Mottle Virus

Peanut mottle virus (PMoV) is a potyvirus with RNA-containing flexuous particles about 740–750 nm long. It is transmitted by a number of aphid species, mechanically, and is seed-borne in peanut, bean, cowpea and two lupin species. Its host range includes cultivated and wild legumes and several other weed species. There are several strains or variants of this virus that can cause a variety of symptoms. Symptoms in bean depend upon varieties; some respond with local chlorotic or necrotic lesions, veinal browning, systemic mottle, apical and stem necrosis followed by death. Others show only local chlorotic or necrotic lesions, and hence are considered to be resistant (Paguio and Kuhn 1973; Bock and Kuhn 1975).

A large number of bean varieties were found to be systemically resistant, since they respond to inoculation with only a localized infection. Inheritance studies have demonstrated that in 'Royalty Purple Pod', resistance is conditioned by a single, but incompletely dominant gene, designated *Pmv* (Provvidenti and Chirco 1987). An effective control of PMoV in bean can be achieved by complete isolation of bean fields from peanut fields, or by planting buffer crops between peanuts and beans. There is a general agreement that infected peanut crops are the major sources of this virus; consequently, only resistant varieties should be considered for areas in which this virus is prevalent (Paguio and Kuhn 1973; Bock and Kuhn 1975).

Soybean Mosaic Virus

Soybean mosaic virus is a potyvirus with long, flexuous virus particles about 750 nm long, containing a single strand of RNA. Serologically it is related to several viruses of the same group. This is one of the most important viruses affecting soybean, in which it is seed- transmitted, hence it can be found wherever this crop is cultivated. In nature, it is spread by aphids, but it can also be easily transmitted mechanically. In bean, the symptoms incited by SMV strongly resemble those caused by strains of bean common mosaic virus, i.e. green mottle, green veinbanding, blistering, downward leaf cupping, and distortion. Certain varieties will respond with systemic necrosis, followed by death (Bos 1972; Costa et al. 1978).

A large number of varieties were determined to be systemically resistant. These develop a visible or invisible local infection, but the virus fails to spread systemically. Whereas in soybean, resistance is strain-specific, the resistance in bean is not. The resistance to SMV in 'Great Northern 1140' is governed by the single incompletely dominant gene *Smv* (Costa and Cupertino 1976; Provvidenti et al. 1982). Avoid planting susceptible bean varieties near soybean fields, which are the major source of the virus. Buffer zones between these two crops can be helpful in preventing infections, and the destruction of soybean volunteers from a previous crop may also eliminate sources of infection.

Watermelon Mosaic Virus

Watermelon mosaic virus is a potyvirus formerly known as watermelon mosaic virus 2 (WMV-2). It is characterized by flexuous particles about 760 nm long that contain a single strand of RNA. WMV is spread by more than 20 aphid species in the nonpersistent manner, and is also easily transmitted by mechanical means, but no seed transmission has been reported for this virus. WMV is able to infect many species in the Cucurbitaceae and Leguminosae, hence it is present in the tropics as well as in temperate and cool regions of the world. Symptoms incited by this virus in bean are similar to those caused by BYMV; however, in addition to the foliar yellow mosaic, pods are often distorted and mottled. Very little is known regarding natural infection and the frequency of this virus in bean crops, because symptoms can be easily confused with those caused by BYMV (Purcifull et al. 1984).

Many varieties appear to be resistant to this virus. In some resistant varieties, there is a local infection, but the virus fails to move systemically. Resistance has been demonstrated to be due to two independently monogenically inherited genes. *Wmv* was found in the variety 'Great Northern 1140' and confers systemic resistance that is temperature-insensitive. *Hsw* was located in the variety 'Black Turtle 1' and confers both local and systemic resistance, but at high temperature (30°C), these plants may develop systemic necrosis, which is conducive to the premature death of infected plants. This reaction appears very similar to the collapse of *II* lines with the necrotic pathotypes of BCMV. A genetic association between *Hsw* and *I* has been reported (Kyle 1988). Pathotype specificity has not been demonstrated for either of these genes (Provvidenti 1974; Kyle and Provvidenti 1987).

Viruses Transmitted by Beetles

This group includes the following viruses: bean curly dwarf mosaic (BCDMV); bean mild mosaic (BMMV); bean pod mottle (BPMV); bean rugose mosaic (BRMV); bean southern mosaic (BSMV); blackgram mottle (BgMV); and cowpea chlorotic mottle (CCMV). They are transmitted by a number of beetle species, including *Cerotoma farialis*, *C. ruficornis*, *C. trifurcata*, *Colaspis flavida*, *C. lata*, *Diabrotica adelpha*, *D. balteata*, *D. undecimpunctata*, *Epilachna varivestis*, *E. vittata*, *Gynandrobrotica variabilis*, and *Paranapiacaba*. After a 24-hour accession feeding, the beetles retain these viruses for 2–3 days. Vectors can acquire these viruses after feeding on infected plants for periods of less than a day and can retain and transmit them up to several days, depending upon the individual viruses (Schultz and Dean 1947; Zaumeyer and Thomas 1948; Thomas 1951; Bancroft 1971; Meiners et al. 1977; Waterworth 1981; Gamez 1982).

Bean Curly Dwarf Mosaic Virus

Bean curly dwarf mosaic virus (BCDMV) is a virus belonging to the cowpea mosaic virus group with particles about 25–28 nm long containing a single-stranded RNA. Serologically it is related to quail pea mosaic. BCDMV is mechanically transmitted and spread by a number of cucumber and Mexican beetles. Originally found in El Salvador, it may be widespread in several South American countries. In addition to common bean, it has been found in other *Phaseolus* species, soybean, pea, chickpea, lentil, broadbean, mung bean, and other leguminous species, wild and cultivated. In bean, an infection at an early stage of growth causes very severe dwarfing. Generally, infected leaves display mosaic, downward curling, rugosity, and twisting. Some varieties will respond with a mild mosaic, whereas others develop apical and stem necrosis and die prematurely (Meiners et al. 1977).

Some bean varieties from Latin America, such as 'Diablo', 'La Vega', 'Santa Ana', and 'Villa Gro', can be considered to be tolerant, because they respond with a mild mottle. These same varieties were found to be tolerant also to bean mild mosaic virus (BMMV); hence, they will also tolerate a dual infection of BCDMV and BMMV (Meiners et al. 1977). Weeds are the major sources of this virus, which is often found in combination with BMMV and other beetle-transmitted viruses. Chemical control and destruction of weeds growing on the edge of fields can minimize infection.

Bean Mild Mosaic Virus

Bean mild mosaic virus (BMMV) is a virus with isometric particles about 28 nm in diameter with a single RNA component. In nature, it is spread by a number of beetles, notably the Mexican bean beetle and the spotted cucumber beetle, but is also transmitted by mechanical means. There is evidence that it may be spread through the soil, but the mechanism is unknown. It infects *Phaseolus* species and many other legumes, but symptoms are veinal chlorosis and chlorotic mottle, which are usually mild and often transitory, since plants tend to recover from these symptoms. When it occurs in association with other viruses, however, synergistic symptoms can be very severe. Serologically it appears unrelated to other isometric viruses, hence BMMV is classified as ungrouped (Waterworth 1981).

Resistance was reported in some wild beans, such as *Phaseolus leptostachyus* and *P. filiformis*. Lima bean (*P. lunatus*) responds with local lesions, without systemic infection (hypersensitive reaction) (Waterworth 1981). A number of bean varieties, although infected, do not show visible symptoms; consequently, detecting the presence of the virus and assessing its spread under field conditions is difficult.

Bean Pod Mottle Virus

Bean pod mottle virus (BPMV) is a comovirus with isometric particles about 28 nm long that include two genome segments of a single-stranded RNA, encapsidated in different particles. The virus is readily transmitted mechanically and spread in nature by several beetle species and can infect bean, other *Phaseolus* spp., soybean, and other cultivated and wild species. In bean, leaf and pod mottle accompanied by distortion are common features of plants infected by this virus. BPMV can severely affect the yield, because of seed abortion (Zaumeyer and Thomas 1948).

A number of bean varieties that are resistant to bean common mosaic virus appear to be resistant to BPMV. Resistant plants respond with brownish local infection, without systemic translocation of the virus. In the varieties 'Pinto 14', 'Great Northern UI No. 1', 'U.S. No. 5 Refugee', and 'Rival', the resistance to systemic infection was determined by Thomas and Zaumeyer (1950) to be conditioned by a single dominant gene. To this gene Provvidenti (1976) assigned the symbol *Bpm*.

Bean Rugose Mosaic Virus

Bean rugose mosaic virus (BRMV) is a comovirus with isometric particles about 28 nm in diameter, containing three components of a single-stranded RNA. It is spread by beetle species and easily transmitted mechanically, but not via seeds. Serologically, it is related to the other comoviruses. The host range is confined mostly to leguminous species. Three strains have been identified that differ in host range but are identical or closely related serologically. Some bean varieties respond with rather severe symptoms, which include very prominent foliar mosaic, rugosity, blistering, and malformation. Pods are also mottled and distorted. Other varieties react with necrotic local lesions without systemic infection, or are completely resistant (Gamez 1982).

Although this virus incites spectacular symptoms in susceptible genotypes, it is of minor importance because a large number of varieties are either hypersensitive resistant (only local necrotic lesions) or are highly resistant. Resistance is monogenic dominant and conferred by three alleles: *Mrf* confers complete resistance and is dominant over Mrf_2. This gene governs hypersensitivity and is dominant over *mrf*, which is responsible for susceptibility. These symbols were given by Machado and Pinchinat (1975). *Mrf*, in Spanish, stands for 'mosaic rugosode feijao' (Machado and Pinchinat 1975).

Bean Southern Mosaic Virus

Bean southern mosaic virus (BSMV) is considered to be the type member of the sobemovirus group, which includes viruses with c. 30 nm isometric particles containing a positive sense, single-stranded RNA. It is distributed worldwide, but it has a host range mostly confined to Leguminosae. Several strains of this virus can be differentiated by diagnostic species. The original strain (type) infects bean, soybean, and lima bean, but not cowpea. The cowpea strain infects cowpea, soybean, pea, and other legumes, but not bean. The Ghana strain infects both bean and cowpea but causes more severe symptoms than the other strains. BSMV is spread by beetles, mechanically, and through seeds (1–5% in bean and 5–40% in cowpea). In bean, the virus incites symptoms of a mild mottle to a prominent mosaic, green veinbanding, and foliar distortion, depending upon the variety (Tremaine and Hamilton 1983).

Some varieties respond with necrotic local lesions without systemic spread (e.g. 'Great Northern 15', 'Great Northern 59', 'Great Northern 123', 'Kentucky Wonder', 'Pinto 78', and 'Red Mexican UI 34'. Zaumeyer and Harter (1943) reported that this resistance is conferred by a single dominant gene, to which Provvidenti (1987) assigned the symbol *Bpm*. Resistance-breaking strains have been reported, but no efforts have been made to find additional sources of resistance. Virus-free seed eliminates the principal source of infection. Chemical control of vectors can be beneficial in some cases.

Blackgram Mottle Virus

Blackgram mottle virus (BgMV) is an RNA-containing virus belonging to the carmovirus group, with virions about 28 nm in diameter. It is seed- and mechanically transmitted, and in nature is spread by several beetle species. It has been reported to occur in India and Thailand, where, in addition to blackgram (*Vigna mungo*), it infects other leguminous species, including common bean. Serologically, BgMV is not related to the other beetle-transmitted viruses. It is seed transmitted in blackgram (c. 5–8%), but no information is available regarding beans. In susceptible bean varieties, the virus causes a variety of symptoms, including foliar mottle, green vein banding, distortion, and plant stunting (Scott and Phatak 1979).

A number of bean varieties, including 'Pinto', respond with necrotic local lesions without systemic infection (Scott and Phatak 1979). This hypersensitivity is a valuable source of resistance, but no information is available regarding its inheritance. The disease can be prevented by not growing bean near blackgram or mungo bean (*Vigna radiata*).

Cowpea Chlorotic Mottle Virus (Bean Yellow Stipple Virus)

Cowpea chlorotic mottle virus (CCMV) is a member of the bromovirus group, with isometric virus particles ranging from 26 to 30 nm in diameter, containing a single strand of RNA. It is disseminated by beetle species and is mechanically transmissible but not spread by seed. It is found in cowpea, bean, soybean, and other cultivated and wild legumes. In bean, initially, some varieties respond with slight malformation of mottled leaves, whereas other varieties, such as 'Great Northern' and 'Pinto', develop the characteristic yellow stippling, which consists of chlorotic spots that tend to coalesce, forming irregularly shaped patches involving as much as three quarters of a leaflet. Subsequent growth usually exhibits very few of these spots. Plants are only slightly stunted, hence, the disease is not destructive (Zaumeyer and Thomas 1957; Bancroft 1971).

It appears that all the varieties tested with this virus were susceptible, but a more intense screening could reveal sources of resistance (Zaumeyer and Thomas 1957).

Viruses Transmitted by Fungi

Tobacco necrosis virus (TNV) is the only member of this group that is able to cause diseases in beans.

Tobacco Necrosis Virus

Tobacco necrosis virus (TNV) is a necrovirus with RNA-containing isometric particles about 26 nm long. Small RNA satellites are associated with this virus that serologically are unrelated but multiply only in plants infected with the virus. All the strains have been classified in two serogroups, A and B. These groups based on serological relationships do not coincide with pathogenicity groups. The virus is readily transmitted mechanically, and spread in nature by zoospores of the chytrid fungus *Olpidium brassicae*, which can rapidly acquire the virus and enter the host's roots. Efficiency of transmission depends upon a suitable combination of strain of the virus, fungus, and the host species. There are indications that the virus does not survive in resting spores of the fungus. The virus is not seed-transmitted, but it survives in dry plant tissue for a long time. It is capable of infecting a large number of plant species, both herbaceous and woody. In a number of species, infection usually remains localized. In common bean, symptoms depend on the virus strain. Some strains are not able to infect this crop, others cause local lesions, whereas certain other strains produce a disease known as Stipple Streak, because of the necrotic streaks along stems and petioles. Systemic symptoms also include foliar necrotic ringspots and yellowing. Pods may also be affected by necrotic rings and deformation. When satellites are present, usually the symptoms are attenuated (Kassanis 1970).

No resistance has been found in varieties of *P. vulgaris*, but one report indicated that lima bean and scarlet runner beans may be resistant. Usually, this virus causes severe infection when beans are grown in fields that have not been utilized for a long period. Beans should not be planted in fields in which this virus has occurred in previous years, because it persists in soil debris (Natti 1959).

Viruses Transmitted by Leafhoppers

Members of this group are beet curly top (BCTV) and tobacco yellow dwarf virus (TYDW), formerly known as bean summer death virus. They are spread by leafhopper species, notably *Circulifer tenellus* and *C. opacipennis*, in which they circulate without multiplying (Thomas and Mink 1979; Thomas and Dowyer 1984).

Beet Curly Top Virus

Beet curly top virus (BCTV) is a geminivirus with isometric particles about 20 nm in diameter, occurring in pairs (geminate) and containing single-stranded DNA. The virus is restricted to the phloem, and is transmissible mechanically only by special procedures. This virus occurs mostly in arid and semi-arid climates, where it can infect a large number of plant species. In bean, the initial symptoms may resemble those caused by bean common mosaic virus, but in later stages the symptoms are very distinct. Laminae of trifoliate leaves are puckered, downward curled, thicker than normal, and very brittle. Affected plants remain severely stunted and often die prematurely; however, the intensity of symptoms varies with the variety, because some are partially tolerant. Ten strains of the virus have been reported. These strains differ in virulence, hence some varieties are resistant or tolerant to one strain and susceptible to others (Thomas and Mink 1979).

The use of resistant varieties can be a very effective control of this virus, which can cause devastating epidemics, particularly in dry beans, due to their long vegetative cycle. Resistance can be strain-specific, however, and some resistant varieties become susceptible if the environmental temperature is too high (>30°C). Schultz and Dean (1947) reported that resistance to BCTV was dominant in 'Common Red Mexican', 'Burthen', and 'Great Northern UI 15'. They also reported that this resistance was due to the interaction of one dominant and one recessive gene, which were inherited independently. Later, the symbol *Ctv* was assigned to the dominant and *ctv-2* to the recessive gene (Provvidenti 1987).

Tobacco Yellow Dwarf Virus (Bean Summer Death)

Tobacco yellow dwarf virus (TYDV) possesses geminate particles (20 nm × 35 nm) containing DNA. Serologically it is distantly related to BCTV. In nature, it is spread by leafhoppers, particularly *Orosius argentatus*, but is not transmissible mechanically or via seed. It is of common occurrence in Australia in tobacco and beans, where it incited a destructive disease known as Bean Summer Death. In bean, the main symptoms consist of down-curling of trifoliates, which develop a diffuse yellowing, causing wilting and premature abscission. As for BCTV, the phloem is affected; consequently, stems develop vascular necrosis and plants may die within ten days after the appearance of the first symptoms, if ambient temperature is high (>30°C) (Thomas and Dowyer 1984).

Ballantyne (1970) in Australia found that varieties resistant or tolerant to BCTV were also resistant to TYDV. These included 'Cascade', 'Picker', 'Prelude', 'Widusa', 'Dark Red Kidney', 'Michelite', 'Seafarer' and others. Hence, adoption of resistant or tolerant varieties can be very beneficial. Whether the same genetic factors are involved in the resistance to BCTV and TYDV is unknown.

Viruses Transmitted by Nematodes

Tobacco ringspot virus (ToRSV) and tomato ringspot virus (TmRSV) are the only members of the group of viruses transmitted by nematodes that have been found to cause diseases in beans. These viruses are acquired by the nematodes within 24 hours and transmitted by adult and larval stages, which retain these viruses for several weeks. Virus particles were found in the lumen of the esophagus of viruliferous nematodes. ToRSV and TmRSV are mainly found in fields that have been placed under cultivation after a prolonged period of rest. Undisturbed soil usually allows the nematodes to multiply abundantly and acquire the virus from endemic plant species (Stace-Smith 1984, 1985).

Tobacco Ringspot Virus

Tobacco ringspot virus (ToRSV) is a nepovirus with particles about 28 nm in diameter with a bipartite RNA genome. In nature it is transmitted by nematodes of the genus *Xiphinema* (*X. americanum*, *X. rivesi*, and other related species). It is easily transmitted mechanically and in seed of some plant species, but not in bean. ToRSV can infect a very large number of cultivated and wild plant species, including woody and herbaceous. In bean, symptoms consist of prominent chlorotic spots formed by concentric rings; some varieties also develop foliar necrosis and may die prematurely. Many varieties tend to recover from the acute stage of infection, however, and during the chronic stage, symptoms are usually mild, although pod production is significantly affected (Stace-Smith 1985).

Tu (1983) located resistance to ToRSV in two navy bean varieties, 'Clipper' and 'Kentwood', and determined that the high level of this resistance is conferred by a single recessive gene. Provvidenti (1987) assigned the symbol *trv* to this gene. This disease may occur in fields that have not been cultivated for several years, if high populations of viruliferous nematodes have developed.

Tomato Ringspot Virus

This nepovirus shares many characteristics with tobacco ringspot virus. Particles are about 28 nm in diameter with a bipartite RNA genome. Serologically, however, these two viruses are not related. In nature, TmRSV is transmitted by nematodes of the genus *Xiphinema* (*X. americanum*, *X. rivesi*, and other related species), and its host range comprises hundreds of plant species, both woody and herbaceous. It is easily transmitted mechanically and in seed of some plant species, but not in bean. Symptoms caused by TmRSV resemble those incited by ToRSV, and hence consist of chlorotic and necrotic mottle with ring-like spots, leaf distortion, and stem and apical necrosis and premature death. A number plants may survive the first stage of infection (acute) and may partially recover from severe symptoms (Stace-Smith 1984).

Varieties resistant to ToRSV are susceptible to TmRSV, and so are hundreds of other varieties of *P. vulgaris* and some related species. One accession of *P. coccineus*, 'Kelvedon Marvel', was highly resistant (Provvidenti 1988b). A good control measure is to avoid planting a bean crop in fields that were uncultivated for a number of years. In some cases, soil fumigation could be used to eliminate the vectors, but it is costly.

Viruses Transmitted by Thrips

Only one virus of this group has been reported to infect beans: tobacco streak virus (TSV).

Tobacco Streak Virus (Bean Red Node)

Tobacco streak virus (TSV) is an ilarvirus with isometric particles ranging from 27 to 35 nm long that contain a single strand of RNA. It can be readily transmitted by mechanical means and in nature is spread by possibly two thrips, *Thrips tabaci* and *Frankliniella occidentalis*. Although TSV is very widely spread, it has only occasionally caused epidemics. It is seedborne in alfalfa, common bean (14%), adzuki bean, chickpea, *Datura*, fenugreek, sweet clover, soybean, cowpea, *Nicotiana clevelandii*, and other species. It can infect many species. In bean, it causes a disease known as Red Node. Recently two pathotypes of this virus were characterized, which can be differentiated by using diagnostic species. Both of these pathotypes can infect bean. Affected plants develop a reddish discoloration of the nodes, accompanied by necrosis and reddening of leaf veins. Necrosis of the stem and lateral shoots may also follow. Sunken, reddish lesions develop on young pods, which may shrivel without forming seeds. Plants infected from seed or in the seedling stage remain very stunted. Pathotype II induces a green to yellow mosaic in some varieties that could be confused with those incited by other viruses (Virgin 1943; Thomas and Zaumeyer 1950; Thomas 1951; Fulton 1985; Kaiser et al. 1991).

Although most of the bean varieties were found to be susceptible to this virus, a few, such as 'Kentucky Wonder', are apparently resistant (Thomas and Zaumeyer 1950). This resistance, however, has not been confirmed by other researchers. Since a relatively high percentage of infection has occurred in bean fields in close proximity to white and yellow sweet clover, fenugreek, alfalfa, and other wild legumes, it is recommended that adequate sanitation measures be developed.

Bean Viruses Transmitted by Whiteflies

Viruses transmitted by whiteflies include the following: bean golden mosaic (BGMV), bean dwarf mosaic (BDMV), euphorbia mosaic (EMV), and rhynchosia mosaic (RMV). Whitefly-transmitted viruses are not as readily transmitted as other viruses, and inoculation efficiency increases with long acquisition periods. In general, these viruses are retained by the vectors for their entire life. Inoculation efficiency increases with vector population size. Major sources of whiteflies are cotton, soybean, and some wild species. These viruses do not multiply in the vectors, nor is there any evidence of transovarial transmission (Bird 1958, 1962; Matyis et al. 1975; Costa and Cupertino 1976; Goodman and Bird 1982; Morales et al. 1990).

Bean Golden Mosaic Virus

Bean golden mosaic virus (BGMV) is a geminivirus with dimer particles about 15–20 nm in diameter. The genome consists of two circular single-stranded DNAs. In nature, this virus is spread by *Bemisia tabaci*, which usually acquire the virus in 3 to 5 hours and may retain it for 20 days or life. Some strains can be transmitted mechanically with ease, others only with great difficulty. BGMV is not seedborne in bean, hence wild legumes represent the major reservoirs. In bean, symptoms are very striking and consist of a bright golden yellow mosaic affecting most of the trifoliates, which are usually deformed and curled downward. Varietal reaction can greatly modify the intensity of these symptoms, however; hence some varieties may be only slightly affected. Plant stunting, pod malformation, discoloration, and reduction of seed size are also common features of the disease (Goodman and Bird 1982).

A massive screening conducted by CIAT (Cali, Colombia) involving more than 10,000 lines failed to reveal a very high level of resistance, but only varying degrees of tolerance in *P. vulgaris*, *P. lunatus*, and *P. coccineus*. This tolerance can be strain-specific; consequently, some varieties appear to be more suitable in one location than in another. The following varieties were reported to be tolerant to BGMV: 'Aete 1-37', 'Aete 1-38', 'Aete 1-40', 'Carioca 99', 'Kranz', 'ICA Pijao', 'Pompsu', 'Porrillo Sintetico', 'Porrillo 780', 'Preto 143-106', 'Rosinha GZ-69', 'ICA Tui', and 'Turrialba 1'. Some of these lines have been used to develop new varieties. Resistance appears to be inherited polygenically. An effort is presently underway to use biotechnology in controlling this very destructive virus. In some cases insecticides, particularly those with systemic action, can be effective in preventing or reducing the spread of the vector and some cultural practices may be beneficial.

Bean Dwarf Mosaic Virus (Bean Chlorotic Mottle)

Bean dwarf mosaic virus (BDMV) is another geminivirus with dimer particles measuring about 20×33 nm containing a single strand of DNA. Serologically it is closely related to BGMV. The vector is *Bemisia tabaci*, which acquires the virus from weed reservoirs, notably *Sida spinosa* and *S. rhombifolia*. BDMV can be transferred mechanically, but the efficiency varies considerably among different varieties. There is no evidence of seed transmission. It infects several plant species, including bean. Symptoms in susceptible bean are characterized by severe plant stunting (dwarfing), foliar distortion, and mosaic, accompanied by bright yellow chlorotic spots (Bird 1958; Morales et al. 1990).

Some varieties respond to infection with a few localized chlorotic spots, hence they are considered to be resistant. These included 'Black Turtle Soup', 'ICA Pijao', 'Pinto 114', 'Redlands Greenleaf B',

'Porrillo Sintetico', 'ICTA-Quetzal', and 'Red Mexican 35'. Some of these varieties are known to be also tolerant to BGMV. No information is available regarding the inheritance of resistance to BDMV (Bird 1958; Morales et al. 1990).

Euphorbia Mosaic Virus

Euphorbia mosaic virus (EMV) is a virus with geminate particles (individual monomer 12–13 nm) containing a single strand of DNA. In some areas of the tropics, it appears of common occurrence in *Euphorbia* spp., from which it spreads to cultivated crops. *Bemisia tabaci*, its vector, acquires the virus after a 10-minute access period, and after 20 minutes of incubation the virus can be transmitted to other plants. It appears to be mechanically transmissible, but there is no indication that the virus is seedborne. It can infect a number of plant species, including lentil, soybean, and other legumes. Symptoms in bean include necrotic spots in the vector feeding areas. Systemic infection usually causes foliar malformation, twisting, and plant stunting. Affected plants tend to develop a number of basal branches, thus giving them a bushy appearance (Matyis et al. 1975; Costa and Cupertino 1976). No resistance has been found in common bean and other *Phaseolus* spp. but it occurs in some varieties of *Vigna angularis*, *V. radiata*, and *V. umbellata* (Matyis et al. 1975; Costa and Cupertino 1976).

Rhynchosia Mosaic Virus

Rhynchosia mosaic virus (RMV) is another whitefly-transmitted virus, also spread by *Bemisia tabaci*. Very little is known about this virus, which occurs also in some areas of the tropics in *Rhynchosia minima*. In common bean, it incites leaf malformation, mottle, and yellowing. Because of the excessive proliferation of basal branches, the plants appear severely stunted (Bird 1958).

Two bean varieties, 'La Vega 19' and 'Santa Ana', are tolerant to the virus and have expressed a significant level of field resistance. Chemical control of the vector and other sanitation measures may limit or reduce the spread of RMV (Bird 1962).

Viruses Transmitted by Unknown Vectors

Tobacco mosaic virus (TMV) is the only virus transmitted by unknown vectors.

Tobacco Mosaic Virus

Tobacco mosaic virus (TMV) is a tobamovirus with particles about 300 × 18 nm. It is easily transmitted mechanically through seeds of some crops but lacks a biological vector. It infects many solanaceous crops systemically, causing economic damage (Zaitlin 1975). In common bean, many varieties are immune, whereas a limited number, such as 'Pinto', 'Black Turtle Soup', 'Scotia', 'Tempo', and others, respond with necrotic local lesions and are useful as assay hosts. A closely related tobamovirus, tomato mosaic virus, does not incite local lesions in beans. Some intermediate strains cause very small pin-point chlorotic spots. The immune reaction was determined to be conferred by a single dominant gene, designated *Tm*, whereas the hypersensitive response in 'Scotia' was monogenic recessive (*tm*) (Thompson et al. 1962).

Conclusion

A wealth of genetic diversity exists in the genus *Phaseolus* that can be tapped to develop lines with improved levels and combinations of disease resistance. Many of the resistances are conditioned by a single allele, the simplest possible case for breeding. With some viruses, such as bean golden mosaic virus, however, massive efforts to identify resistance have failed to identify any genes for high levels of resistance. These situations compel us to look to other methods to provide the resistance necessary to stabilize production of this important staple crop, especially in tropical and subtropical areas. Laboratory-based methods will provide a useful complement to the genetic resources already identified in the development of disease-resistant crop varieties.

Literature Cited

Ali, M. A. 1950. Genetics of resistance to the common bean mosaic virus (Bean Virus 1) in the bean (*Phaseolus vulgaris*). Phytopathology 40:69–79.

Alconero, R., and J. P. Meiners. 1974. The effect of environment on the response of bean cultivars to infection by strains of bean common mosaic virus. Phytopathology 64:679–682.

Ashby, J. W. 1984. Bean leaf roll virus. Descriptions of Plant Viruses No. 286. Kew, Surrey, England: Commonw. Mycol. Inst./Assn. Appl. Biol.

Ballantyne, B. 1970. Field reactions of bean varieties to summer death in 1970. Plant Dis. Rep. 54:903–905.

Bancroft, J. B. 1971. Cowpea chlorotic mottle virus. Descriptions of Plant Viruses No. 49. Kew, Surrey, England: Commonw. Mycol. Inst./ Assn. Appl. Biol.

Barnett, O. W., J. W. Randles, and P. M. Burrows. 1987. Relationships among Australian and North American isolates of the bean yellow mosaic potyvirus group. Phytopathology 77:291–299.

Bird, J. 1958. Infectious chlorosis of *Sida carpinifolia* in Puerto Rico. Univ. Puerto Rico Agr. Exp. Station Tech. Pap. 22:1–35 Bird, J. 1962. A whitefly-transmitted mosaic of *Rhynchosia minima* and its relation to tobacco leaf curl and other virus diseases of plants in Puerto Rico. Phytopathology 52:286.

Bird, J., J. Sanchez, R. L. Rodriguez, A. Cortes-Monllor, and W. J. Kaiser. 1974. A mosaic of beans (*Phaseolus vulgaris* L.) caused by a strain of common cucumber mosaic virus. J. Agr. Univ. Puerto Rico 53:151–161.

Bock, K. R., and M. Conti. 1974. Cowpea aphid-borne mosaic virus. Descriptions of Plant Viruses No. 134. Kew, Surrey, England: Commonw. Mycol. Inst./Assn. Appl. Biol.

Bock, K. R., and C. W. Kuhn. 1975. Peanut mottle virus. Descriptions of Plant Viruses No. 141. Kew, Surrey, England: Commonw. Mycol. Inst./Assn. Appl. Biol.

Bos, L. 1970. Bean yellow mosaic virus. Descriptions of Plant Viruses No. 40. Kew, Surrey, England: Commonw. Mycol. Inst./Assn. Appl. Biol.

Bos, L. 1972. Soybean mosaic virus. Descriptions of Plant Viruses No. 93. Kew, Surrey, England: Commonw. Mycol. Inst./Assn. Appl. Biol.

Bos, L., and D. Z. Maat. 1974. A strain of cucumber mosaic virus seed transmitted in bean. Neth. J. Plant Path. 80:113–123.

Brick, M. A., M. H. Dickson, G. C. Emery, S. Magnuson, and H. F. Schwartz. 1990. *Phaseolus* Crop Advisory Committee 1989 Update. Bean Improv. Coop. Annu. Rep. 33: xi–xxiv.

Costa, C. L., and F. P. Cupertino. 1976. Avaliacao das perdas na producao do feijoeiro causadas pelo virus do mosaico dourato. Fitopatol. Bras. 1:18–25.

Costa, A. S., N. Lima, V. da Costa, A. L. D'Artagnan, and E. Bulisani. 1978. Suscetibilidade de certos grupos de feijoerio a infeccao sistemica pelo virus do mosaico comun da soja. Fitopatol. Bras. 3:27–37.

Davis, R. F., and R.O Hampton. 1986. Cucumber mosaic virus isolates seed-borne in *Phaseolus vulgaris*: Serology, host-pathogen relationships, and seed transmission. Phytopathology 76:999–1004.

Dickson, M. H., and J. J. Natti. 1968. Inheritance of resistance of *Phaseolus vulgaris* to bean yellow mosaic virus. Phytopathology 58:1450.

Drijfhout, E. 1978. Genetic interaction between *Phaseolus vulgaris* and bean common mosaic virus with implications for strain identification and breeding for resistance. Agric. Res. Rept. 872, Pudoc, Wageningen, Holland: Centre for Agricultural Publishing and Documentation.

Drijfhout, E., and L. Bos. 1977. The identification of two new strains of bean common mosaic virus. Neth. J. Plant Pathol. 83:13–25.

Drijfhout, E., M. J. Silbernagel, and D. W. Burke. 1978. Differentiation of strains of bean common mosaic virus. Neth. J. Plant Pathol. 84:13–26.

Francki, R. I. B., D. W. Mossop, and T. Hatta. 1979. Cucumber mosaic virus. Descriptions of Plant Viruses No. 213 (revised). Kew, Surrey, England: Commonw. Mycol. Inst./Assn. Appl. Biol.

Fulton, R. W. 1985. Tobacco streak virus. Descriptions of Plant Viruses No. 307 (revised). Kew, Surrey, England: Commonw. Mycol. Inst./Assn. Appl. Biol.

Gamez, R. 1982. Bean rugose mosaic virus. Descriptions of Plant Viruses No. 246. Kew, Surrey, England: Commonw. Mycol. Inst./Assn. Appl. Biol.

Goodman, R. M., and J. Bird. 1982. Bean golden mosaic virus. Descriptions of Plant Viruses No. 102. Kew, Surrey, England: Commonw. Mycol. Inst./Assn. Appl. Biol.

Grogan, R. G., and J. C. Walker. 1948. A pod-distorting strain of the yellow mosaic virus of bean. J. Agr. Res. 77:301–314.

Hedrick, U. P., W. T. Tapley, G. P. Van Eseltine, and W. D. Enzie. 1931. The vegetables of New York Vol. 1, Part II. Beans of New York. Geneva, NY: New York State Agr. Exp. Sta. Publication. 110 p.

Hoch, H. C., and R. Provvidenti. 1978. Ultrastructural localization of bean common mosaic virus in dormant and germinating seeds of *Phaseolus vulgaris*. Phytopathology 68:327–330.

Hollings, M., and O. M. Stone. 1974. Clover yellow vein virus. Descriptions of Plant Viruses No. 131. Kew, Surrey, England: Commonw. Mycol. Inst./Assn. Appl. Biol.

Hubbeling, N. 1972. Resistance in beans to strains of bean common mosaic virus. Mededelingen van de Faculteit Landb. Rijk Gent 28:1025–1033.

Jasper, E. M. J., and L. Bos 1980. Alfalfa mosaic virus. Descriptions of Plant Viruses No. 229 (revised). Kew, Surrey, England: Commonw. Mycol. Inst./Assn. Appl. Biol.

Kaiser, W. J., S. D. Wyatt, and R. E. Klein. 1991. Epidemiology and seed transmission of two tobacco streak virus pathotypes associated with seed increases of legume germplasm in eastern Washington. Plant Dis. 75:258–264.

Kassanis, B. 1970. Tobacco necrosis virus. Descriptions of Plant Viruses No. 14. Kew, Surrey, England: Commonw. Mycol. Inst./Assn. Appl. Biol.

Kline, R. E., S. D. Wyatt, and W. J. Kaiser. 1988. Incidence of bean common mosaic virus in USDA *Phaseolus* germplasm collection. Plant Dis. 73:301–302.

Kyle, M. M., and R. Provvidenti. 1987. Inheritance of resistance to potato Y viruses in *Phaseolus vulgaris* L: 1. Two independent genes for resistance to watermelon mosaic virus 2. Theor. Appl. Genet. 74:595–600.

Kyle, M. M. 1988. The *I* gene and multiple virus resistance in *Phaseolus vulgaris* L. Cornell University, Ph.D. Thesis.

Machado, P. F. R., and A. M. Pinchinat. 1975. Herencia de la reaction de frijol comune a la infection por el virus del mosaic rugoso. Turialba 25:418–419.

Matyis, J. C., D. M. Silva, A. R. Oliveira, and A. S. Costa. 1975. Purificacao e morfologia do virus do mosaico dourado do tomaterio. Summa Phytopath. 1:267–274.

Meiners, J. P., H. E. Waterworth, R. H. Lawson, and F. F. Smith. 1977. Curly dwarf mosaic virus of beans from El Salvador. Phytopathology 67:163–168.

Morales, F. J., and L. Bos. 1988. Bean common mosaic virus. Descriptions of Plant Viruses No. 337. Kew, Surrey, England: Commonw. Mycol. Inst./Assn. Appl. Biol.

Morales, F., A. Niessen, B. Ramirez, and M. Castano. 1990. Isolation and partial characterization of a geminivirus causing bean dwarf mosaic. Phytopathology 80:96–101.

Natti, J. J. 1959. A systemic disease of beans caused by tobacco necrosis virus. Plant Dis. Rep. 43:640–644.

Paguio, O. R., and C. W. Kuhn. 1973. Strains of peanut mottle virus. Phytopathology 63:976–980.

Petersen, H. J. 1958. Beitrage zur Genetik von *Phaseolus vulgaris* L auf Infektion mit Phaseolus Virus 1 Stamm Voldagsen. Zeitschrift Planz. 39:187–224.

Pierce, W. H. 1934. Viroses of the bean. Phytopathology 24:87–115.

Pierce, W. H. 1935. The identification of certain viruses affecting leguminous plants. J. Agr. Res. 51:1017–1039.

Provvidenti, R. 1974. Inheritance of resistance to watermelon mosaic virus 2 in *Phaseolus vulgaris*. Phytopathology 64:1448–1450.

Provvidenti, R. 1976. Reaction of *Phaseolus* and *Macroptilium* species to a strain of cucumber mosaic virus. Plant Dis. Rep. 60:289–293.

Provvidenti, R. 1987. List of genes in *Phaseolus vulgaris* for resistance to viruses. Bean Improv. Coop. Annu. Rep. 30:1–4.

Provvidenti, R. 1988a. Inheritance of resistance to broadbean wilt virus in bean. HortScience 23:895–896.

Provvidenti, R. 1988b. Reaction of bean cultivars to tomato ringspot virus. Bean Improv. Coop. 31:144–145.

Provvidenti, R., and E. M. Chirco. 1987. Inheritance of resistance to peanut mottle virus in *Phaseolus vulgaris*. J. Heredity 78:402–403.

Provvidenti, R., A. C. Monllor, C. L. Niblett, J. Bird, K. H. Gough, 1992. Host differentiation of potyviruses infecting passionfruit (*Passiflora edulis*) in Puerto Rico and Australia. Phytopathology 82:610 (abstr.)

Provvidenti, R., D. Gonsalves, and P. Ranalli. 1982. Inheritance of resistance to soybean mosaic virus in *Phaseolus vulgaris.* J. Heredity 73:302–303.

Provvidenti, R., D. Gonsalves, and M. A. Taiwo. 1983. Inheritance of resistance to blackeye cowpea mosaic and cowpea aphid-borne mosaic viruses in *Phaseolus vulgaris*. J. Heredity 74:60–61.

Provvidenti, R., and W. T. Schroeder. 1969. Three heritable abnormalities of *Phaseolus vulgaris*: Seedling wilt, leaf-rolling, and apical chlorosis. Phytopathology 59:1550–1551.

Provvidenti, R., and W. T. Schroeder. 1973. Resistance in *Phaseolus vulgaris* to the severe strain of bean yellow mosaic virus. Phytopathology 63:196–197.

Provvidenti, R., M. J. Silbernagel, and W. Y. Wang. 1984. Local epidemic of NL-8 strain of bean common mosaic virus in bean fields of western New York. Plant Dis. 68:1092–1094.

Purcifull, D., and D. Gonsalves. 1985. Blackeye cowpea mosaic virus. Descriptions of Plant Viruses No.. 305. Kew, Surrey, England: Commonw. Mycol. Inst./Assn. Appl. Biol.

Purcifull, D. E., E. Hiebert, and J. Edwardson. 1984. Watermelon mosaic virus 2. Descriptions of Plant Viruses No. 293. Kew, Surrey, England: Commonw. Mycol. Inst./Assn. Appl. Biol.

Reddick, D., and V. B. Stewart. 1919. Transmission of the virus of bean mosaic in seed and observations on thermal death-point of seed and virus. Phytopathology 9:445–450.

Schroeder, W. T., and R. Provvidenti. 1970. Resistance of bean (*Phaseolus vulgaris*) to the PV2 strain of bean yellow mosaic virus conditioned by the single dominant gene *By*. Phytopathology 60:1312–1313.

Schultz, H. K., and L. L. Dean. 1947. Inheritance of curly top disease reaction in bean, *Phaseolus vulgaris*. J. Am. Soc. Agron. 39:47–51.

Schwartz, H. F., and M. A. Pastor-Corrales, Eds. 1989. Bean Production Problems in the Tropics, 2nd ed. Cali, Colombia: CIAT.

Scott, H. A., and H. C. Phatak. 1979. Properties of blackgram mottle virus. Phytopathology 69:345–348.

Stace-Smith, R. 1984. Tomato ringspot virus. Descriptions of Plant Viruses No. 290 (revised). Kew, Surrey, England: Commonw. Mycol. Inst./Assn. Appl. Biol.

Stace-Smith, R. 1985. Tobacco ringspot virus. Description of Plant Viruses No. 309 (revised). Kew, Surrey, England: Commonw. Mycol. Inst./Assn. Appl. Biol.

Stewart, V. B., and D. Reddick. 1917. Bean mosaic. Phytopathology 7:61.

Taylor, R. H., and R. S. Greber. 1973. Passionfruit woodiness virus. Descriptions of Plant Viruses No. 122. Kew, Surrey, England: Commonw. Mycol. Inst./Assn. Appl. Biol.

Taylor, R. H., and L. L. Stubs. 1972. Broadbean wilt virus. Descriptions of Plant Viruses No. 81. Kew, Surrey, England: Commonw. Mycol. Inst./Assn. Appl. Biol.

Tatchell, S. P., J. R. Baggett, and R. O. Hampton. 1985. Relationship between resistance to severe and type strains of bean yellow mosaic virus. J. Am. Soc. Hort. Sci. 110:96–99.

Thomas, H. R., and W. J. Zaumeyer. 1950. Inheritance of symptom expression of pod mottle virus. Phytopathology 40:1007–1010.

Thomas, H. R., and W. J. Zaumeyer. 1950. Red node, a virus disease of beans. Phytopathology 40:832–846.

Thomas, J. E., and J. W. Dowyer. 1984. Tobacco yellow dwarf virus. Description of Plant Viruses No. 278. Kew, Surrey, England: Commonw. Mycol. Inst./Assn. Appl. Biol.

Thomas, P. E., and G. I. Mink. 1979. Beet curly top virus. Descriptions of Plant Viruses No. 210. Kew, Surrey, England: Commonw. Mycol. Inst./Assn. Appl. Biol.

Thomas, W. D. 1951. Seed transmission of red-node virus in beans. Phytopathology 41:764–765.

Thompson, A. E., R. L. Lower, and H. H. Thornberry. 1962. Inheritance in beans of the necrotic reaction to tobacco mosaic virus. J. Heredity 63:89–91.

Tremaine, J. H., and R. I. Hamilton. 1983. Bean southern mosaic virus. Description of Plant Viruses No. 274 (revised). Kew, Surrey, England: Commonw. Mycol. Inst./Assn. Appl. Biol.

Tu, J. C. 1983. Inheritance in *Phaseolus vulgaris* cv. 'Kentwood' of resistance to a necrotic strain of bean yellow mosaic virus and to a severe strain of tobacco ringspot virus. Can. J. Plant Path. 5:34–35.

Uyemoto, J. K., and R. Provvidenti. 1974. Isolation and identification of two serotypes of broad bean wilt virus. Phytopathology 64:1547–1548.

Virgin, J. 1943. An unusual bean disease. Phytopathology 33:743–745.

Wade, B. L., and W. J. Zaumeyer. 1940. Genetic studies of resistance to alfalfa mosaic virus and stringiness in *Phaseolus vulgaris*. J. Amer. Soc. Agron. 32:127–134.

Wang, W. Y., G. I. Mink, and M. J. Silbernagel. 1982. Comparison of direct and indirect enzyme-linked immunosorbent assay (ELISA) in the detection of bean common mosaic virus. Phytopathology 72:954.

Waterworth, H. 1981. Bean mild mosaic virus. Descriptions of Plant Viruses No. 231. Kew, Surrey, England: Commonw. Mycol. Inst./Assn. Appl. Biol.

Whipple, O. C., and J. C. Walker. 1941. Strains of cucumber mosaic virus pathogenic on bean and pea. J. Agr. Res. 62:27–60.

Zaitlin, M. 1975. Tobacco mosaic virus. Descriptions of Plant Viruses No. 151. Kew, Surrey, England: Commonw. Mycol. Inst./Assn. Appl. Biol.

Zaumeyer, W. J., and L. L. Harter. 1943. Inheritance of symptom expression of bean mosaic virus 4. J. Agr. Res. 67:295–300.

Zaumeyer, W. J., and H. R. Thomas. 1948. Pod mottle, a virus disease of beans. J. Agr. Res. 77:81–96.

Zaumeyer, W. J., and H. R. Thomas. 1957. Yellow stipple, a virus disease of beans. Phytopathology 40:847–859.

CHAPTER 7

Genetics of Broad Spectrum Viral Resistance in Bean and Pea

MOLLY M. KYLE & ROSARIO PROVVIDENTI

Introduction

Breeding for resistance to plant viruses and plant viral disease is an important objective for many vegetable improvement programs. Relatively few alternatives for direct control of crop losses from viral diseases exist, and genetic resistance is by far the most effective, efficient, and durable strategy presently available. Despite the economic importance of genes for viral resistance in plants, very little is known about how these genes are organized in the host genome or how they function to restrict viral replication and development of symptoms (Fraser 1986, 1987, 1990; Ponz and Bruening 1986). For the purposes of this paper, *resistance* is defined as the ability of a plant to limit both expression of disease and multiplication of the pathogen. This definition excludes tolerant responses, the condition whereby a plant grows normally or nearly normally but supports viral replica-

Dr. Kyle is in the Department of Plant Breeding and Biometry, Cornell University, Ithaca, NY 14853. Dr. Provvidenti is in the Department of Plant Pathology, New York State Agricultural Experiment Station, Cornell University, Geneva, NY 14456. This work was supported in part by USDA/NRI/CGP Grant No. 9101626 and by the Burroughs Wellcome Fund Fellowship of the Life Sciences Research Foundation (MMK).

tion. Tolerance to the virus may be synonymous with resistance to the disease caused by the virus. This distinguishes genes that prevent viral replication and long distance movement (i.e. infection) from host responses that reduce symptom severity and thus losses to disease but that permit viral replication and may not impede spread of the pathogen.

Viral resistance may show dominance, recessiveness, additive and/or epistatic genetic components. Although many genes for viral resistance characterized thus far function monogenically, even simply inherited resistance factors can be subject to complex interactions with modifying factors. Genotype, environment, and interactive genotype × environment effects can be very significant in determining the type(s) of symptoms expressed and severity of losses. Finally, epidemiological factors must be considered in identifying and prioritizing breeding objectives.

The inherent difficulties of breeding for any complex genetic characteristic are clearly evident in the development of advanced virus-resistant breeding lines. Symptoms of viral disease can resemble other physiological and pathological conditions and can be highly variable even among uniformly susceptible populations as a function of host genotype, viral isolate, environmental conditions, and a number of other factors. Several classes of symptom-modulating small RNA molecules may further complicate reproducibility when present in viral cultures. Foliar symptoms observed in greenhouse seedling tests may or may not reflect ability to yield a marketable product in the field. Still, subjective evaluation of the intensity of symptoms, both at the seedling stage and throughout the growing season, remains the primary method of selection in most breeding programs. Easily quantified parameters such as virus titer do not necessarily correlate with ability to yield after inoculation and are often too expensive, time-consuming, and cumbersome for assaying large numbers of samples. Because viruses are obligate pathogens, inoculum cannot be produced *in vitro* and transmission of certain plant viruses requires introduction by a vector.

Most vegetable crops are infected by a number of viruses, including several that regularly cause significant losses across years and locations. Although many viruses may plague a crop species throughout its range of cultivation, regional differences in importance are common. When resistant lines are introduced, reductions in yield and/or quality often continue caused by viruses previously considered of secondary importance (Munger 1993, this volume). Thus, multiple viral resistance has become a major objective in many vegetable crops, especially tomato, pepper, lettuce, legumes, and cucurbit species. Usually, each disease problem is handled separately in breeding programs because of the commonly accepted generalization that monogenic resistance, which is the most simple to transfer, is narrow in effect, i.e. isolate- or strain-specific. When advanced lines are developed, resistance to each pathogen must be combined and selected in a commercially acceptable type through backcrossing or intercrossing. For transfer of recessive genes and oligogenic or polygenic characters, this final phase may require a number of generations. Clearly, simply inherited multiple disease resistance would expedite the development of varieties with adequate levels of resistance to several viral diseases. The existence of genes capable of conditioning resistance to more than one virus, however, has been largely discounted (Fraser 1986).

This chapter summarizes genetic studies in bean and pea that demonstrate the existence of genes and gene families that confer simply inherited resistance to a broad spectrum of plant viruses. Although our results are among the first conclusively to establish patterns of association among viral resistance genes, these data are entirely consistent with both evolutionary and biochemical predictions. Additional examples of genes and gene clusters for multiple viral resistance will certainly be identified upon systematic investigation, particularly among those genes that affect the outcome of infection by related pathogens. The arrangement and relationships among genes for viral resistance may shed light on the dynamic evolutionary processes driven by the host-virus interaction that result in both the existence and organization of these genes within the plant genome, and may define advantageous systems for the elucidation of mechanisms of viral pathogenesis and host resistance.

The *I* Gene in *Phaseolus vulgaris*

The *I* gene for resistance to bean common mosaic virus (BCMV) was first identified in the generally susceptible bean variety 'Stringless Green Refugee' in 1931 by Corbett at the Sioux City Seed Co. in

Sioux City, Iowa (Pierce 1934). Because of the severity and ubiquity of this virus, the *I* gene has been used extensively in bean breeding since its discovery, and a wealth of germplasm has been developed, including diverse resistant and susceptible near-isogenic lines. Under most conditions, the dominant allele confers both local and systemic resistance and eliminates seed transmission (Ali 1950). The virus cannot be recovered from inoculated tissue, although subliminal infection in which single inoculated cells contain the virus has not been conclusively ruled out. No BCMV pathotypes capable of producing mosaic on *I/I* or *I/i* genotypes have been identified, but under certain conditions BCMV may incite very rapid and lethal systemic necrosis. The mechanism that accounts for the relationship between conditional veinal necrosis and resistance to BCMV is not understood, but based on the necrotic phenotype, resistance conferred by the *I* gene has been classified as hypersensitive.

BCMV is one of at least 175 definitive and possible members of the potyvirus group, which accounts for approximately 35% of all known plant viruses (Milne 1988). This group, named for the type member potato virus Y, is typified by a flexuous rod encapsidating a single-stranded monopartite RNA genome translated as a polyprotein. The group is considered to be the most destructive assemblage of plant viruses and is characterized by extensive interrelationship among its members (Hollings and Brunt 1981). Often a number of potyviruses infect a single crop species (Edwardson 1974). All the viruses involved in reports of broad spectrum resistance in bean, pea, and other crops, which have not been so thoroughly investigated, belong to this grouping of pathogens.

The necrotic response to BCMV in *I/-* genotypes (i.e. *I/I* and *I/i*) without modifying genes is apparent within 2–4 days under appropriate conditions and can be induced in three ways: (1) with elevated temperature (>30°C) after inoculation, (2) by approach-graft inoculation with a susceptible infected host at any temperature, or (3) by inoculation with the temperature-independent necrotic BCMV pathotypes NL-3, NL-5, or NL-8 (Drijfhout 1978). The term pathotype is used here to denote an isolate or set of isolates with a unique pattern of resistance and susceptibility on a set of host differentials. The necrotic reaction generally begins with dark pinpoint lesions on inoculated leaves. Expression of the response in detached tissue permits rapid nondestructive screening for the allele (Kyle and Dickson 1988). In whole plants, necrosis extends through the phloem into the major veins, petiole, and stem, culminating in death of the apical meristem in 4–7 days. This rapid and distinctive necrosis was first noted in the field and termed *black root*. Eventually, BCMV was identified as the causal agent (Grogan and Walker 1948). In the last several years, necrosis-inducing isolates have caused severe losses in Africa, the Caribbean basin, parts of the North American Midwest, and most recently in the Northwest seed production areas, with the result that breeding strategies and the usefulness of existing varieties have been reconsidered (Van Rheenen and Muigai 1984; Provvidenti 1990). Lethal necrosis from BCMV is still relatively rare in the field, however, and the *I* allele remains widely used. Subtle differences in expression of this response are apparent in different genetic backgrounds, and there are additional genes that limit the necrotic response to inoculated leaves.

Four unlinked loci designated *bc-u*, *bc-1*, *bc-2*, and *bc-3* comprise a second independent gene system for systemic resistance to mosaic symptoms incited by BCMV (Drijfhout et al. 1978). Allelic series occur at *bc-1* and *bc-2* (Drijfhout 1978). Resistance conditioned by these loci is pathotype-specific and depends upon cooperation between *bc-u/bc-u* and one or more of the *bc-x* loci in the homozygous recessive condition. In resistant plants, the virus replicates in inoculated leaves but fails to move systemically. In some heterozygous combinations, viral replication is substantially reduced, although systemic movement still occurs (Day 1984). A host differential series has been developed that distinguishes the eleven viral virulence groups identified thus far (Drijfhout et al. 1978). Before describing the interaction between oligogenic, recessive, pathotype-specific resistance conferred by the *bc* loci, and the monogenic, dominant, hypersensitive resistance conditioned by *I*, it is useful to contrast these two gene systems for BCMV resistance in the common bean.

Despite wide deployment of the *I* gene for decades in all continents where beans are cultivated and the extensive pathogenic diversity of BCMV, no isolate has been identified that is capable of inciting the mosaic reaction on *I/-* genotypes. Apparently, the *I* gene precludes expression of the mosaic response to all BCMV isolates tested by us or others to date, although several closely related potyviruses can cause mosaic on both *I/-* and *i/i* lines. When the *I* gene is overcome and the virus moves systemically, movement is always accompanied by rapid death of the vascular tissue. All BCMV isolates tested thus far, including representatives of Drijfhout's "non-necrotic" pathotypes, are capable of inducing the necrotic reaction under certain conditions on *I/-* genotypes (Kyle and Provvidenti 1987a). In contrast, the *bc* gene system for oligogenic recessive resistance is probably the most complex example of strain-specific or pathotype-specific viral resistance described to date (Fraser 1986). The two gene systems in *P. vulgaris* operate independently. In *I/-* genotypes that also carry *bc* alleles, the *bc* system effectively overlays the reaction determined at the *I* locus and protects against systemic lethal necrosis.

Provvidenti et al. (1983) assigned two gene symbols, *Cam* and *Bcm*, for dominant resistance in the common bean to two potyviruses related to BCMV, cowpea aphid-borne mosaic virus (CAMV) and blackeye cowpea mosaic virus (BlCMV). Interestingly, the response to these potyviruses conditioned by *Cam* and *Bcm* was indistinguishable from that of *I/-* genotypes when inoculated with BCMV (Kyle 1988). A fourth potyvirus, watermelon mosaic virus (WMV) (formerly watermelon mosaic virus-2), also incited temperature-sensitive necrosis on lines that carried a dominant resistance factor (Kyle and Provvidenti 1987b). Currently studies are underway that suggest dominant resistance to one isolate of a fifth potyvirus, passionfruit woodiness virus (PWV), is associated with the *I* gene (R. Provvidenti, unpublished observation). Finally, linkage was observed in bean lines known to carry the *I* allele with temperature-independent veinal necrosis incited by a sixth potyvirus, soybean mosaic virus (SMV) (M. M. Kyle and R. Provvidenti, in press). This dominant lethal response was also controlled by a single gene.

The relationship between dominant, hypersensitive, temperature-dependent resistance to BCMV, BlCMV, CAMV, WMV and the lethal dominant response to SMV was investigated. PWV was not included in this study as its involvement in this association had not been discovered. All genotypes evaluated that included a number of lines representing very diverse germplasm showed an all-or-none pattern of response to the potyviruses. The variety 'Corbett Refugee', in which the original putative spontaneous mutation at *I* occurred, failed to develop mosaic with any of the five viruses, although the line in which this mutation was identified, 'Stringless Green Refugee', was uniformly susceptible. This observation was true for all *I/I* lines tested, including a number that were unrelated to 'Corbett Refugee' and therefore presumably a consequence of an independent mutational event at this locus (Kelly 1988). A large test for linkage between responses to the five potyviruses failed to recover any recombinants among 1,000 segregating F_3 families. Expanded linkage formulae were developed to estimate a maximum limit for recombination frequency (M. M. Kyle and S. J. Schwager, in preparation). If distinct factors account for the observations, they must be extremely tightly linked (Kyle 1988). Any genetic background that suppressed or otherwise modified expression of necrosis with BCMV similarly altered expression of necrosis with the other four viruses, thereby indicating that the very complex epistatic interactions involved in the production of this phenotype were conserved across viruses and genotypes. This genetic block of hypersensitivity to five potyviruses was closely linked to the *B* locus for darkened testa color approximately 25 units from *St*, stringy pod, in Lamprecht's linkage group III (Kyle and Dickson 1988).

These results in the common bean establish that broad spectrum resistance to plant viral disease can be simply inherited and handled as a single genetic unit in breeding programs. Data suggest that a single allele at the *I* locus confers hypersensitivity to five related potyviruses, BCMV, BlCMV, CAMV, SMV and WMV. It is possible, however, that tandem duplication has occurred, creating tightly linked, distinct loci, or a pseudoallelic series at a complex locus. Thus far, exhaustive attempts have failed to identify any evidence of multiple factors at or near the *I* locus (M. M. Kyle et al., in preparation). Detailed evolutionary studies of other types of gene systems suggest that all three of the genetic configurations that could account for simply inherited multiple plant virus resistance (single alleles, pseudoallelic series, and series of distinct but very tightly linked loci) will eventually be identified.

Viral Resistance Conferred by Gene Clusters in *Pisum sativum*

Recent findings in *P. sativum* have provided further examples of multiple potyvirus resistance, in this case, conferred by two very tightly clustered arrays of recessive genes located on different chromosomes. For years, it had been noted anecdotally by breeders that domestic varieties resistant to bean yellow mosaic virus (BYMV) were also resistant to other potyviruses such as pea mosaic virus (PMV), clover yellow vein virus (CYVV), the lentil strain of pea seed-borne mosaic virus (PSbMV-L1), and two of the bean viruses discussed above, the NL-8 pathotype of bean common mosaic virus (BCMV) and watermelon mosaic virus (WMV) (e.g. Schroeder and Provvidenti 1971). The close relationship of these six viruses and co-segregation of resistance during breeding suggested a genetic association. The discovery of a pea line from China (PI 391530), resistant to BYMV and WMV but susceptible to the other viruses, implied that resistance was governed by more than one genetic factor. When a collection of plant introductions and populations developed by intercrossing were screened, five distinct linked loci for recessive resistance to six potyviruses were identified.

Four of the five genes in the cluster on chromosome 2, *bcm*, *cyv-1*, *pmv*, and *sbm-2*, function independently of ambient temperature to confer complete systemic resistance to BCMV NL-8, CYVV, PMV, and PSbMV-L1, respectively (Provvidenti 1987; Provvidenti 1990a; Provvidenti and Alconero 1988a). Resistance to BYMV and WMV is controlled by the fifth gene, *mo*, which is temperature-sensitive in the heterozygous state (Marx and Provvidenti 1979). At 15°C, inoculated *mo*/+ plants remain symptomless, whereas at 28°C or above, a prominent mosaic develops (Schroeder and Provvidenti 1971). Selection of this gene cluster on chromosome 2 can be expedited by using the isozyme locus *Pgm-p* (phosphoglucomutase) as a marker (Weeden et al. 1984).

A second cluster of recessive genes has been identified in plant introductions from North India and Ethiopia. Five tightly linked genes on chromosome 6, *sbm-1*, *sbm-3*, *sbm-4*, *wlv*, and *cyv-2*, are responsible for resistance to three PSbMV strains, standard, lentil, P4, white lupine mosaic virus (WLMV) and CYVV, respectively

(Hagedorn and Gritton 1973; Provvidenti 1987; Provvidenti and Alconero 1988b; R. Provvidenti, unpublished data). Previous research has established that *sbm-1* is approximately 4 map units from *wlo* (wachlos), a recessive gene for the waxless condition of leaves and stipules (Gritton and Hagedorn 1975). With this morphological character as a marker, it has been possible to transfer resistance to PSbMV and CYVV for six generations without viral tests. The pair of linked genes *sbm-3* and *cyv-2* on chromosome 6, and *sbm-2* and *cyv-1* on chromosome 2, may be repetitive entities, since they independently confer resistance to the same isolates of each virus (Provvidenti 1987; Provvidenti and Alconero 1988b).

Gene clusters such as those found in pea have been termed *linkats* by Demarly (1979). In the course of evolution, ancestral genes may have duplicated and given rise to tandem arrays that in some instances may have translocated to other chromosomes. This hypothesis could account for the observed linkages on chromosomes 2 and 6, and may explain the presence of linked genes with identical pathotype specificity for resistance to CYVV and PSbMV-L1 on both chromosomes. Conversely, it could be hypothesized that loci for resistance to pea viruses may have originated as independent mutations in different linkage groups and converged on chromosomes 2 and 6.

Conclusions

The two cases described above constitute the most thoroughly investigated examples of simply inherited multiple plant viral resistance. Although the genetic basis that accounts for these observations differs in bean and in pea, many of the implications of this discovery apply regardless of genetic fine structure. Co-segregation of genes for viral resistance has been observed by plant breeders over the last fifty years and noted at least indirectly in several studies (e.g. Cook 1960; Cockerham 1970; Provvidenti et al. 1983). Three generalizations consistent with our work in bean and pea emerge from the literature. First, in these cases resistance to each virus is inherited identically. Second, the resistant reaction to each virus is phenotypically identical. Third, the viruses involved are members of the potyvirus group.

It has been reported that dominant resistance to three potyviruses, potato viruses A, C, and Y, now considered two viruses, PVA and PVY, is conditioned by one gene in some diploid *Solanum* species allied with potato, including *S. chacoense* (Cockerham 1970). Recessive resistance to PVY, pepper mottle virus (PeMV) and tobacco etch virus (TEV) was associated in *Capsicum* species (Cook 1960; Zitter and Cook 1973). Finally, there is evidence from cucumber breeding programs in which Chinese accessions have served as the source of incompletely dominant or recessive resistance to three potyviruses (W. Meijsing, H. M. Munger, personal communication). Resistance to WMV appears to be associated with similar resistance to zucchini yellow mosaic virus (ZYMV) in segregating populations, although the genetic basis of this association appears complex (R. Grumet, personal communication). Resistance to these two viruses also may be associated with incompletely dominant resistance to a third cucurbit potyvirus, papaya ringspot virus (PRV), in some genetic backgrounds (W. Meijsing, personal communication). This possibility cannot be ruled out in the Cornell program.

Resistance to WMV and BCMV is associated in both bean and pea. Although these two viruses are somewhat related serologically and are identical in particle morphology, they differ considerably with respect to biological properties. WMV has a very broad host range, including at least 160 species in 23 dicotyledonous families, is not transmitted through seed or pollen in legumes, and has been assigned to subgroup III based on large, virally encoded proteinaceous structures (inclusion bodies) that appear in both the cytoplasm and nucleus of infected cells (Purcifull and Hiebert 1984). In contrast, BCMV is transmitted through seed and pollen in *P. vulgaris* and has a narrow host range primarily limited to legumes (Bos 1971). Only cytoplasmic inclusions are observed in BCMV-infected tissue, possibly indicative of significant differences in genome structure between WMV and BCMV. Thus, there are clear biological and structural differences between these two viruses, and the mechanisms of resistance are entirely unrelated in the two host species, bean and pea. However, the identity or tight linkage of resistance to BCMV and WMV in bean and pea is suggestive of some relationship or shared aspect(s) of pathogenicity between this pair of viruses not revealed by conventional criteria. The determination of affinities within the potyvirus family has proven difficult as assessments of structural similarities differ with methods applied (Harrison 1985) and biochemical data or assessments based on nucleotide sequence do not necessarily correlate well with clear differences in biological properties (e.g. Shukla and Ward 1989; Van der Vlugt et al. 1989). The existence of simply inherited multiple resistance may provide clues to relationships among pathogens based on functional rather than structural similarities.

In light of the extensive adaptive radiation that has resulted in an extraordinary continuum of pathogenic diversity within the potyvirus group, it will be interesting to discover whether the phenomenon of simply inherited broad spectrum resistance is limited to resistance against this group of plant viruses. Efforts are continuing to identify genes and gene clusters for multiple plant viral resistance in several host species to determine the extent to which these results can be generalized to streamline breeding efforts and to identify patterns of association among members of the potyvirus group. These studies to map viral resistance genes in crop genomes may define systems in which fundamental questions of evolution of host resistance and plant viral pathogenicity and stability of viral resistance can be approached.

Our results suggest an additional important parameter in the characterization of genes for resistance to plant viruses, namely the concept of the resistance spectrum of a gene or gene cluster, essentially the converse of the host range of a pathogen (Kyle 1988). This attribute may be applied in the description of viral resistance genes and may be potentially useful in identifying relationships among the viruses affected by the same or similar host genes. Overlapping resistance spectra of genes from different host species suggest relationships between viruses not revealed by other criteria. These relationships may indicate specific molecular and biochemical similarities in the process of plant viral pathogenesis, its interruption, and evolution. Conserved viral processes and gene products provide potential targets for engineering broad spectrum resistance and may also identify specific regions for mutational analysis to identify determining viral components of the virus-host interaction.

In conclusion, our work has clearly established the phenomenon of simply inherited resistance to multiple plant viruses. This discovery, based on systematic analysis of the organization of plant viral resistance genes in the host genome, has direct implications for both conventional and molecular strategies for crop improvement. Immediate considerations for conventional breeding extend both to the organization and management of segregating populations and to the selection of sources of resistance to be employed. Information derived from fundamental studies of these systems could provide specific conventional and molecular approaches to generate novel sources of viral resistance derived from the viral and/or host genome, and expand our understanding of how resistance genes function to limit viral infection.

Literature Cited

Ali, M. A. 1950. Genetics of resistance to the common bean mosaic virus in the bean (*Phaseolus vulgaris* L.). Phytopathology 40:69–79.

Bos, L. 1971. Bean common mosaic virus. Descriptions of Plant Viruses No. 73. Kew, Surrey, England: Commonw. Mycol. Inst./Assn. Appl. Biol.

Cockerham, G. 1970. Genetical studies on resistance to potato viruses X and Y. Heredity 25:309–348.

Cook, A. A. 1960. Genetics of resistance in *Capsicum annuum* to two virus diseases. Phytopathology 50:364–367.

Day, K. L. 1984. Resistance to bean common mosaic virus in *Phaseolus vulgaris*. Ph.D. Thesis, University of Birmingham, England.

Demarly, Y. 1979. The concept of linkat. In Proc. Conf. Broadening Genet. Base Crops, Wageningen, 1978. Pudoc, Wageningen, Holland.

Drijfhout, E. 1978. Genetic interaction between *Phaseolus vulgaris* L. and bean common mosaic virus and its strains. Agric. Res. Rep. 872, Pudoc, Wageningen, Holland: Centre for Agricultural Publishing and Documentation.

Drijfhout, E., M. J. Silbernagel, and D. W. Burke. 1978. Differentiation of strains of bean common mosaic virus. Netherlands J. Plant Pathol. 84:13–26.

Edwardson, J. R. 1974. Host ranges of the Potato Virus Y family. Fla. Agric. Expt. Sta. Mono. Ser. No. 5.

Fraser, R. S. S. 1986. Genes for resistance to plant viruses. CRC Crit. Rev. Plant Sci. 3:257–294.

Fraser, R. S. S. 1987. Biochemistry of Virus-Infected Plants. New York: John Wiley & Sons, Inc.

Fraser, R. S. S. 1990. The genetics of resistance to plant viruses. Ann. Rev. Phytopathol. 28:179–200.

Gritton, E. T., and D. J. Hagedorn. 1975. Linkage of the pea genes *sbm* and *wlo*. Crop Sci. 15:447–448.

Grogan, R. G., and J. C. Walker. 1948. The relation of common mosaic to black root of beans. J. Agric. Res. 77:315–331.

Hagedorn, D. J., and E. T. Gritton. 1973. Inheritance of resistance to the pea seed-borne mosaic virus. Phytopathology 63:1130–1133.

Harrison, B. D. 1985. Usefulness and limitations of the species concept for plant viruses. Intervirology 24:71–78.

Hollings, M., and A. A. Brunt. 1981. Potyviruses. In Handbook of Plant Virus Infections and Comparative Diagnosis. Ed. E. Kurstak. New York: Elsevier/North Holland Biomedical Press.

Kelly, J. D. 1988. Is there more than one source of the *I* gene? Bean Improvement Coop. Ann. Rep. 31:148–149.

Kyle, M. M. 1988. The *I* gene and broad spectrum plant virus resistance in *Phaseolus vulgaris* L. Ph.D. Thesis, Cornell University, Ithaca, NY.

Kyle, M. M., and M. H. Dickson. 1988. Linkage of hypersensitivity to five potyviruses with the *B* locus for seed coat color in *Phaseolus vulgaris* L. J. Heredity 79:308–311.

Kyle, M. M., and R. Provvidenti. 1987a. A severe isolate of bean common mosaic virus NY 15. Bean Improvement Coop. Ann. Rept. 30:87–88.

Kyle, M. M., and R. Provvidenti. 1987b. Inheritance of resistance to Potato Y viruses in *Phaseolus vulgaris* L. I. Two independent genes for resistance to watermelon mosaic virus 2. Theor. Appl. Genet. 74:595–600.

Kyle, M. M., and R. Provvidenti. Inheritance of resistance to potyviruses in *Phaseolus vulgaris* L. II. Linkage relations and utility of a dominant gene for lethal systemic necrosis to soybean mosaic virus. Theor. Appl. Genet. in press.

Marx, G. A., and R. Provvidenti. 1979. Linkage relations of *mo*. *Pisum* Newsletter 11:28–29.

Milne, R. G. 1988. The Plant Viruses, Vol. 4. The Filamentous Plant Viruses. New York: Plenum Press.

Munger, H. M. 1992. Breeding for viral disease resistance in cucurbits. In Resistance to Viral Diseases of Vegetables: Genetics and Breeding. Ed. M. M. Kyle. Portland, OR: Timber Press.

Pierce, W. H. 1934. Viroses of the bean. Phytopathology 24:87–115.

Ponz, F., and G. Bruening. 1986. Mechanisms of resistance to plant viruses. Ann. Rev. Phytopathol. 24:355–381.

Provvidenti, R. 1987. Inheritance of resistance to clover yellow vein virus in *Pisum sativum*. J. Heredity 78:126–128.

Provvidenti, R. 1990a. Inheritance of resistance to pea mosaic virus in *Pisum sativum*. J. Heredity 81:143–145.

Provvidenti, R. 1990b. Reactions of some leading bean cultivars to African and indigenous strains of bean common mosaic virus. Bean Improvement Coop. Ann. Rep. 33:167–168.

Provvidenti, R., and R. Alconero. 1988a. Inheritance of resistance to a lentil strain of pea seed-borne mosaic virus in *Pisum sativum*. J. Hered. 79:45–47.

Provvidenti, R., and R. Alconero. 1988b. Inheritance of resistance to a third pathotype of pea seed-borne mosaic virus in *Pisum sativum*. J. Hered. 79:76–77.

Provvidenti, R., D. Gonsalves, and M. A. Taiwo. 1983. Inheritance of resistance to blackeye cowpea mosaic and cowpea aphid-borne mosaic viruses in *Phaseolus vulgaris*. J. Hered. 74:60–61.

Purcifull, D. E., and E. Hiebert. 1984. Watermelon mosaic virus–2. Descriptions of Plant Viruses No. 293. Kew, Surrey, England: Commonw. Mycol. Inst./Assn. Appl. Biol.

Schroeder, W. T., and R. Provvidenti. 1971. A common gene for resistance to bean yellow mosaic virus and watermelon mosaic virus 2 in *Pisum sativum*. Phytopathology 61:846–848.

Van der Vlugt, R., S. Allefs, P. de Hann, and R. Goldbach. 1989. Nucleotide sequence of the 3′-terminal region of potato virus Y^N RNA. J. Gen. Virol. 70:229–233.

Van Rheenen, H. A., and S. G. S. Muigai. 1984. Control of bean common mosaic by deployment of the dominant gene *I*. Netherlands J. Pl. Pathol. 90:85–94.

Weeden, N. F., R. Provvidenti, and G. A. Marx. 1984. An isozyme marker for resistance to bean yellow mosaic virus. J. Heredity 75:411–412.

Zitter, T. A., and A. A. Cook. 1973. Inheritance of tolerance to a pepper virus in Florida. Phytopathology 63:1211–1212.

CHAPTER 8

Application of Genetic Theory in Breeding for Multiple Viral Resistance

BRIAN T. SCULLY & WALTER T. FEDERER

Introduction

Among the broad array of disease agents that affect crop plants, viruses are potentially some of the most devastating. Because viruses are obligate parasites and are often insect-vectored, the methods used to control these diseases differ from the methods used to to control fungal and bacterial diseases. Viral diseases are commonly managed in three ways: regulatory measures, pesticides targeted at vectors or alternative hosts, and genetic resistance. Regulatory controls such as indexing, crop-free periods, and the elimination of alternate hosts have proven useful but are not guaranteed. Pesticides can augment the regulatory process by reducing vector and alternate host populations, although this is expensive, often not effective, and raises environmental concerns. In nature, genetic resistance provides the best defense against plant disease, and this mechanism remains the safest and most economical approach for crop protection (Browning 1980). *Genetic resistance* is defined as any heritable

Dr. Scully is at the Everglades Research and Education Center, University of Florida, Belle Glade, FL 33430. Dr. Federer is in the Biometrics Unit, Department of Plant Breeding and Biometry, Cornell University, Ithaca, NY 14853. Florida Agriculture Experiment Station Journal Series No. R02305.

trait that reduces the effect of the virus (Russell 1978). The highest level of resistance results from a lack of recognition between the pathogen and host and the subsequent inability of the virus to reproduce in the host (Gracen 1982). Tolerance provides a lower and less acceptable level of resistance, primarily because the virus replicates in the host organism. Tolerant plants show less severe disease symptoms, less damage to the economic organs, and/or suppression of yield by the pathogen (Russell 1978). Morphological traits that limit a vector's ability to transmit a virus mechanically also contribute to host plant resistance.

To develop genetically resistant varieties, resistant germplasm must first be identified. Natural populations provide the original genetic resource from which to extract, identify, and characterize the genes that confer resistance. Locally adapted and foreign varieties are initially screened for resistance genes, followed by land races and wild accessions from the crop's center of diversity or origin. Because crop and disease often co-evolve, genetically diverse forms of resistance are likely to be available from one of these germplasm sources. If the desired level of resistance is not found within the species of interest, methods such as gene manipulation or transfer from related species are required (see Superak et al. 1993).

Once resistant genotypes are identified, the inheritance of resistance must be determined. The mode of inheritance and the crop's reproductive biology establish the selection technique and breeding method most likely to maximize genetic gain from selection. Oligogenically inherited traits are more easily handled than polygenic traits, which require the use of more complex selection procedures (Mayo 1987). The choice of a selection procedure must also be guided by the genetic variability of the pathogen (Day 1974; Simmonds 1979). A basic understanding of viral epidemiology, as well as the mechanisms of resistance and pathogenicity, is useful in the development of resistant varieties. Ultimately, the virus and variety with the resistance genes must interact freely in a cropping system. The final success of a resistant variety is judged by its productivity under field conditions.

Our purpose in this chapter is to compare selection techniques and breeding methods useful for the development of resistant varieties. We review successful breeding methods and present new permutations of these schemes. Breeding for resistance to one disease is relatively straightforward; the difficulty arises in breeding for multiple viral resistance (Khush 1980; Provvidenti 1985). Flexibility in the choice and application of breeding methods is needed to "pyramid" resistant genes into breeding lines and varieties quickly and easily (Nelson 1973). We apply the general theory of selection for multiple quantitative traits to selection procedures for multiple viral resistance. Multiple trait selection techniques are associated with a number of possible breeding methods, thus these concepts are developed for the transfer of mono/oligogenically inherited traits in autogamous diploid species. Additionally, we present an appropriate set of equations for the determination of population or sample size. An effective population size maximizes the probability of successfully finding a desired genotype, without demanding excessive resources.

Selection Techniques

Incorporation of resistance to a single virus is accomplished via selection over one or more generations. In contrast, breeding for multiple viral resistance presents three other possible approaches to selection, including tandem selection, independent culling levels, and the selection index (Turner and Young 1967; Baker 1986). These techniques were developed primarily for quantitative traits and are used regularly in most plant and animal breeding programs; they also provide the theoretical foundation for the selection of qualitatively inherited traits.

In tandem selection, a single trait is selected discretely for one or more generations until the desired phenotype is obtained. Selection is then applied for a second, third or more traits in the same way. For mono/oligogenically inherited traits, a single generation of selection should suffice to obtain the desired phenotype, provided the population size is adequate. Homozygous genotypes are subsequently confirmed in a progeny test generation. With independent culling levels, two or more traits are selected concurrently over every generation of selection until the desired phenotype is obtained. Like tandem selection, independent culling levels can be applied to either oligogenic or polygenic traits. Variations in these two selection methods include selection for a set of multiple traits jointly in a single generation, and concurrent selection for subsets of two or more traits with each subset in a different generation. The selection index was developed specifically for quantitative traits (Smith 1936). It requires the use of genetic variances, covariances, and economic weights to rank and select individuals. Qualitatively inherited traits are occasionally incorporated into modified selection index models as categorical traits (Van Vleck 1979).

In breeding for multiple viral resistance, the biology of the different viruses and their interaction with the host species affect the choice of a selection technique. Cross protection, synergism, variations in symptomatology and escapes influence selection for multiple viral resistance. Cross protection resulting from simultaneous inoculations of closely related viruses can confound identification of resistant and susceptible phenotypes, and affect the accuracy of selection. In addition, the interaction of different viruses with the different genetic backgrounds of the host may cause symptomology to vary, becoming an unreliable criteria for selection. Synergism among viruses may require simultaneous inoculations with two or more viruses to identify resistance to a viral complex. When synergism exists, inoculation with the viral complex and the single viruses can clarify their pathogenicity and interaction. Tandem selection removes the problem of cross protection but fails to address synergism, although resistance to a viral complex can be treated as a distinct trait. Inoculation of genetically different plants with a single virus does not remove possible differences in symptomology, but it does clarify the variation. Lastly, escapes affect all selection techniques, but can be detected with repeat inoculations within a generation or over successive generations.

No particular technique for multiple trait selection is best for the development of multiple viral resistant genotypes. The three approaches presented above are flexible; perhaps variations or different combinations of these techniques in different generations would yield the desired result. Tandem selection is the easiest, simplest, and the least complicated, under most circumstances.

Breeding Methods

In self-pollinated crops, the development of disease-resistant varieties is routinely accomplished with the backcross and pedigree methods (Allard 1960). Backcross methods rapidly introgress specific resistance genes from a donor parent into a desirable variety, concurrently reconstructing the original variety. A new variety is produced quickly, requires minimal testing and is nearly isogenic with the original. Backcross methods are particularly desirable when the donor parent is unadapted or genetically distant from the original variety. When disease problems arise unexpectedly, the backcross method is the method of choice.

Pedigree methods such as F_2 selection, single seed descent (Brim 1966; Empig and Fehr 1971), and the nested hierarchy (Cockerham 1954) are also useful for incorporating disease resistance. Pedigree methods are useful when both parents are adapted and carry genes for resistance to different diseases. Compared with the backcross, a more genetically diverse germplasm base is developed, but the time required to release a variety is much longer. In a pedigree method, recombination and segregation among the unselected genes permits the development of unique genotypes and phenotypes. The backcross and pedigree methods are forms of inbreeding that ultimately result in the development of pure lines. The level of homozygosity (H) increases with each generation of selfing or backcrossing, and is defined as

$$H = \{(2^g - 1) \ / \ (2^g)\}^l \qquad 1.$$

where g is the number of backcrossed or selfed generations (for F_2, $g=1$; F_3, $g=2$, etc.); and l is the number of loci under selection (Allard 1960). The frequency of the desirable homozygous genotypes in any selfed generation is defined as $0.5H$, in the absence of linkage.

Backcross Methods

There are numerous permutations of the backcross method; each addresses a given breeding objective. These permutations are divided into three categories according to the inheritance and number of traits under transfer. For resistance to a single virus conditioned by a dominant gene, the simple backcross is the standard method (Figure 1). When resistance to a single virus is conferred by a recessive gene, the alternate backcross and self (Figure 2), continuous backcross (Figure 3), and the simultaneous backcross and self (Figure 4) methods are acceptable. For the transfer of resistance to multiple viruses, methods such as the sequential backcross (Figure 5), parallel backcross (Figure 6), and the multiple trait backcross (Figure 7) are effective.

The simple backcross procedure transfers a dominant gene from a single donor parent (D_1) to the original variety or recurrent parent (R) (Figure 1). The F_1 generation is composed of resistant heterozygous individuals. The backcross generations are composed of susceptible homozygous recessive and heterozygous resistant individuals at expected frequencies of ½ each. The heterozygous resistant individuals are tested for, and selected in each backcross generation ($BC_1 \rightarrow BC_v$), then crossed to the recurrent parent. Heterozygous individuals are selected in the final backcross generation and selfed to produce homozygous and heterozygous resistant, and homozygous susceptible progeny at expected genotypic frequencies of ¼, ½, and ¼, respectively. Homozygous and heterozygous resistant individuals from this generation are confirmed and separated in the last generation (Figure 1).

The alternate backcross and self procedure transfers a recessive allele from the donor parent (D_1) to the recurrent parent (R) (Figure 2). In this case, each backcross generation is composed of homozygous and heterozygous susceptible individuals at expected genotypic frequencies of ½ each. A selfed generation is included after each backcross generation to reveal the recessive individuals. If more than 5 ($n \geq 5$) backcross individuals are selfed to produce a backcross F_2 generation, there is a $\geq$ 95% (P_α) chance that at least one individual is a heterozygous carrier of the recessive allele (Table 1). Backcross F_2 families derived from each of the backcross individuals are tested for resistance, and a recessive individual is selected for crossing to the recurrent parent. Homozygous recessive individuals are confirmed in the last generation (Figure 2).

The continuous backcross method transfers a resistant gene without the inclusion of a selfing generation (Figure 3). It serves for the transfer of a recessive allele masked in the heterozygous condition, but is not strictly limited to this case. In this method, the

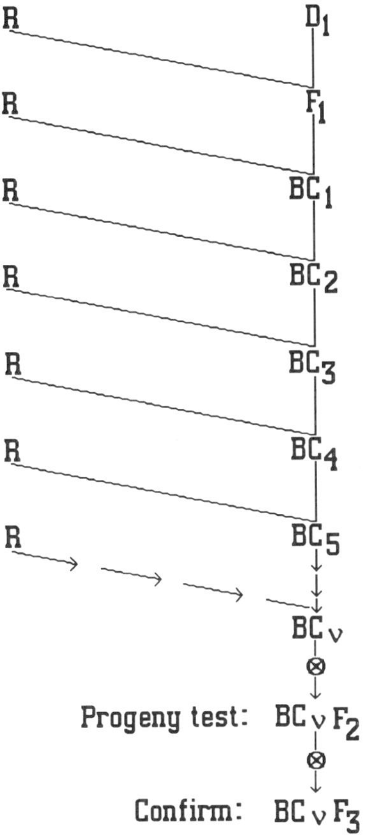

Figure 1. *The simple backcross procedure transfers a dominant gene from a donor parent (D_1) to the recurrent parent (R). Individuals are tested and selected in each backcross generation (BC_1 to BC_ν) and then progeny tested and confirmed in the selfed generations (BC_ν F_2 and BC_ν F_3).*

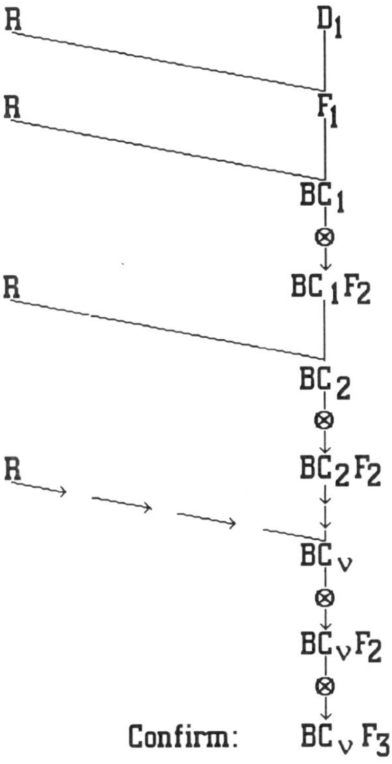

Figure 2. *The alternate backcross and self procedure incorporates a recessive gene from a donor parent (D_1) to a recurrent parent (R). Progenies are tested and selected in each backcross selfed generation (BC_1 F_2 to BC_ν F_2) and the gene is confirmed in the last generation (BC_ν F_3).*

identification and selection of the desired genotype is delayed until the end of the breeding program. To compensate for the absence of selection, the number of crosses and population size is increased. If *n* individuals from the BC_1 generation are crossed to the recurrent parent, the probability that any one of these individuals is a heterozygous or homozygous equals ½. For a 95% chance of randomly selecting a heterozygote from a cross in BC_1, n equals 5 plants. To maintain this probability through repeated backcross generations, n more individuals are derived from each of the previous generation's n individuals. Thus, the number of plants increases exponentially as n, $n^2, n^3, n^4, \ldots n^{\nu}$. By BC_5, n^5 or 3125 individuals are required to maintain a probability of success (P_{α}) at 95%. As the number of individuals increases exponentially (n^{ν}), the resources required also increase compared to the alternate backcross and self procedure. More realistically, only a few backcross generations are done continuously before selfing to uncover the recessive gene. The process is again repeated until the trait is transferred. This approach is convenient when a few backcross generations are produced in the off season and the selfed generation planted for selection.

The simultaneous backcross and self method (Figure 4) is a permutation of the backcross that is used primarily to transfer a recessive gene, but it requires less labor and space than the continuous backcross. Like the continuous backcross, the number of individuals (n) used for the crosses in each generation determines the probability that a heterozygous individual is chosen. In every backcross F_2 generation, the crosses with the heterozygous individuals are revealed and the homozygous crosses discarded. If partial, over-, or co-dominant gene action controls the expression of a trait, then the heterozygous individuals are more easily distinguished and these recessive gene methods may be unnecessary (see Munger 1993).

The primary advantage of the continuous backcross and the simultaneous backcross and self method is that a breeding objective is attained much faster than is possible with the alternate backcross and self procedure (Figures 2, 3, and 4). Additionally, more crosses and the larger population sizes in these backcross schemes provide greater opportunity to break linkage, recover the original phenotype quickly, and find exceptional recombinants. There is a greater likelihood of these events in the continuous backcross because the number of crosses increases by the power of n compared to the simultaneous backcross and self scheme in which n remains constant.

The sequential backcross is a set of simple backcross cycles run in

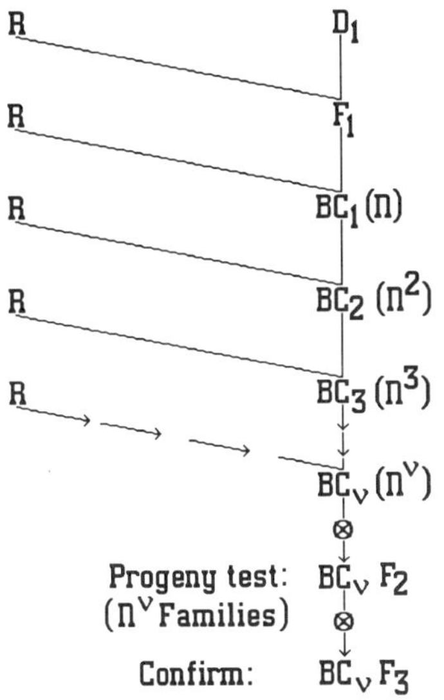

Figure 3. *The continuous backcross procedure primarily transfers a recessive gene. Initially,* n *plants in the BC_1 generation are crossed to the recurrent parent (R) and crosses are increased exponentially* (n^2, n^3,...n^v) *with each backcross generation. Progeny testing, selection, and confirmation of the trait are delayed until the last two generations (BC_v F_2 and BC_v F_3).*

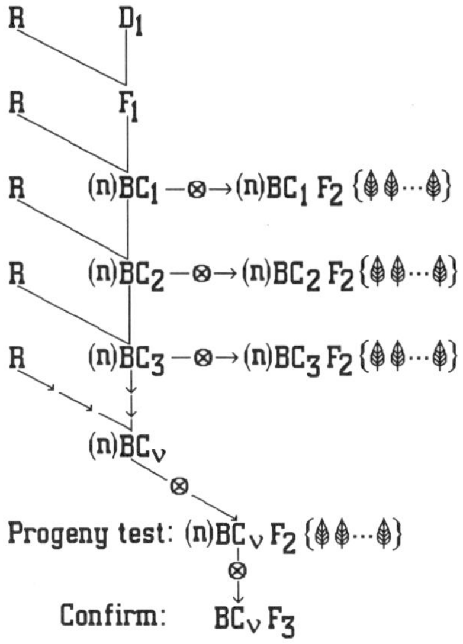

Figure 4. *The simultaneous backcross and self method is primarily used to transfer a recessive gene from a donor parent (D_1) to a recurrent parent (R). A number of plants* (n) *are each crossed to the recurrent parent and concurrently selfed to produce an F_2, and to uncover the recessive gene. A backcross parent with the recessive gene is selected for crossing in the next generation.*

succession with each trait incorporated separately over time (Figure 5). From different donor parents (D_1, D_2, D_3, etc.), dominant or recessive genes can be transferred. Selection is practiced in the backcross generations for dominant genes, but recessive genes require a selfing generation. At the end of each cycle, a new donor parent is incorporated into the breeding program and the line or variety derived from the previous backcross cycle becomes the recurrent parent. The process repeats with as many donor parents as are needed. Progeny tests can be performed after each backcross cycle and prior to the introduction of a new donor parent, or after all traits are incorporated. This method is the slowest of the multiple trait backcrosses but is commonly used to deploy resistance genes for newly discovered diseases.

The parallel backcross technique is a set of simple backcrosses performed concurrently (Figure 6). Donor parents, each with a different dominant gene, are backcrossed to the same recurrent parent until the desired phenotype is obtained. At some point in the scheme,

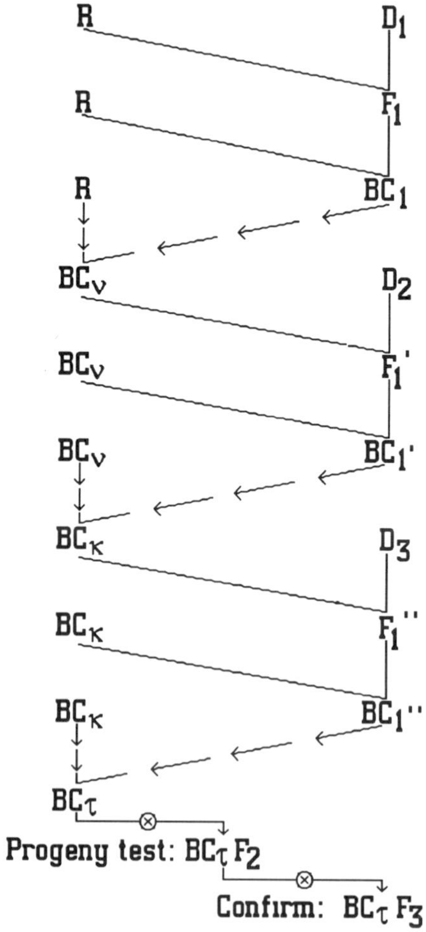

Figure 5. *The sequential backcross procedure is used to incorporate dominant genes, each from a different donor parent (D_1, D_2, etc.).*

individuals with the desired trait from each donor are mated to produce the F_1' generation. The F_1' generation is used to produce the double cross F_1'' generation, which is selfed to produce the F_2 (Figure 6). Putative homozygous individuals are identified in the F_2, and F_3 families are produced. These F_2-derived F_3 families are parti-

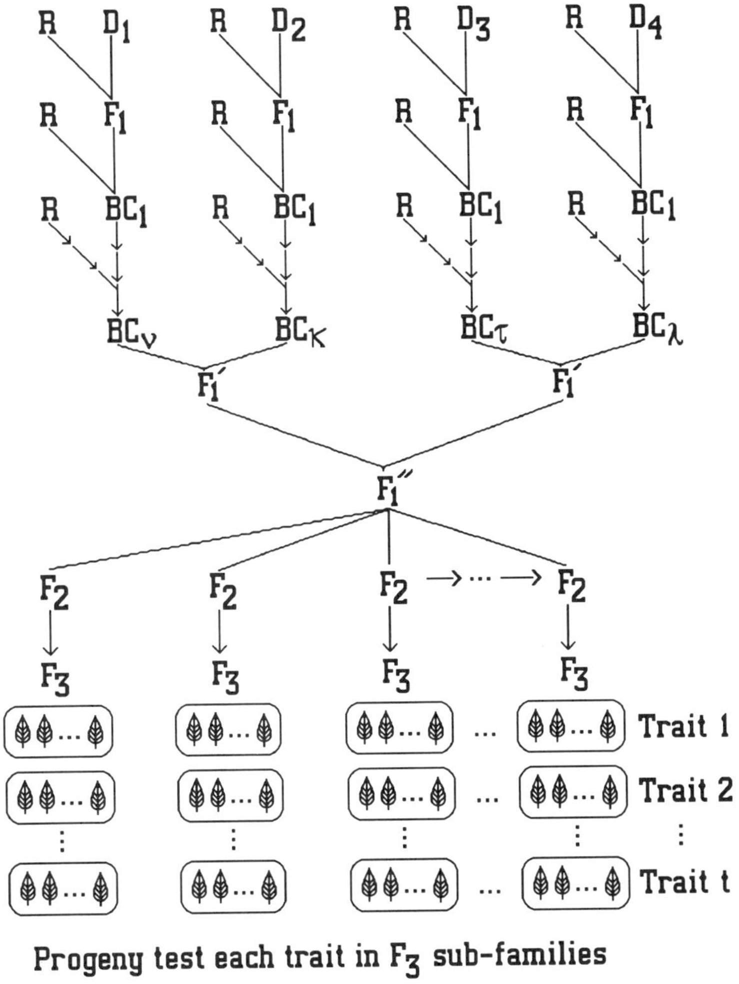

Figure 6. *The parallel backcross scheme primarily incorporates different dominant genes concurrently, each from a different donor parent (D_1 to D_4) into the same recurrent parent (R).*

tioned into subfamilies, and homozygosity is confirmed for each trait (Figure 6). If viruses cross protect, multiple trait selection is delayed until the F_3.

In the parallel and sequential backcross methods (Figures 5 and 6), single traits are initially transferred independently and selected discretely in each backcross generation. In the parallel backcross, all traits are ultimately combined and selected together in the F_3 generation. Both methods are presented for the transfer of dominant genes, but can be modified for the transfer of recessive genes. In the parallel backcross, selfing is integrated into the basic procedure. Insertion of selfing generations between the last backcross and the F_1' generations, and between the F_1' and the F_1'' generations, permits the identification and selection of the recessive genotypes. Throughout the parallel backcross, the recurrent parent is consistent and the coancestry of the progenies converge on the recurrent parent. In the sequential backcross, the recurrent parent changes with each cycle of backcrosses, and the progenies genetically diverge from the original recurrent parent as the number of cycles becomes large.

In the multiple trait backcross (Figure 7), genes are transferred jointly and selected in a single generation. The number of traits transferred is limited by the donor parent and the population size. As the number of genes under transfer increases, the population size must increase commensurately. This method is an extension of the simultaneous backcross and self method (Figure 4), except that the n backcrossed F_2 families are partitioned into subfamilies and evaluated for each trait. Partitioning is not needed for morphological traits, but is advised for identification of viral resistance genes. Unlike the sequential or parallel backcross, no modification of this method is required to handle recessive genes.

All of these backcross methods are generalized schemes that can be modified or combined to meet breeding objectives; all are subject to the usual assumptions assigned to the backcross (Allard 1960; Simmonds 1979). Every trait should have easily distinguishable classes (qualitative distribution), and the genes that condition a trait should be highly penetrant (Suzuki et al. 1985). The single trait methods provide a framework upon which the multiple trait backcross methods are constructed. With the multiple trait methods,

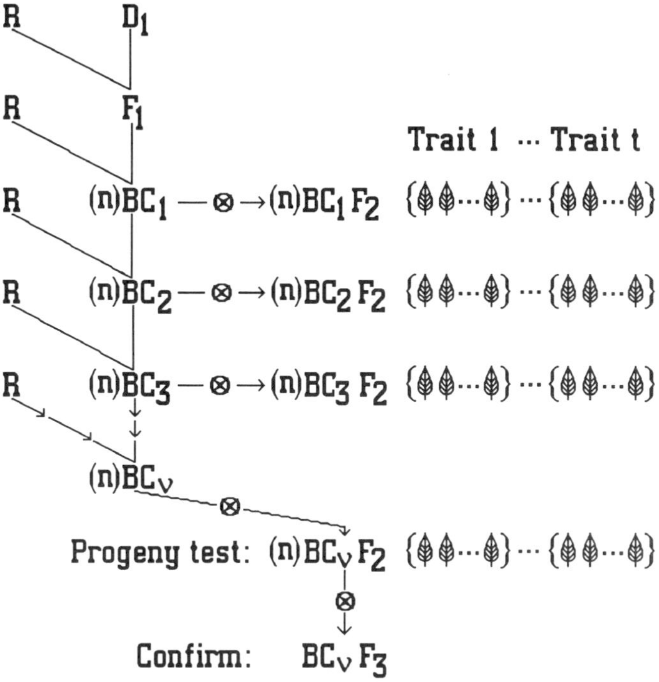

Figure 7. *The multiple trait backcross method transfers two or more traits from the same donor parent* (D_1)*. A number of plants* (n) *are crossed to the recurrent parent (R) and concurrently selfed and tested for each trait. The backcross parent with each trait is selected for crossing in the next generation.*

greater resources and record keeping are required, but the breeding goal is accomplished more quickly. The expression of a trait prior to flowering is desirable in certain applications of the backcross, but irrelevant in the recessive gene methods (Figures 2, 3, and 4) or the multiple trait backcross (Figure 6). In many viral resistance breeding programs, inoculation is done early in the plant life cycle and susceptibility is determined before flowering. Use of both backcross and simultaneous self methods (Figures 4 and 7) assumes multiple flowers, although maize (*Zea mays* L.) breeding techniques allow the same female inflorescence to be selfed and outcrossed (Sheridan and Clark 1987).

The continuous, parallel, and the simultaneous self and backcross methods were developed and refined by Henry M. Munger and identified as "Munger's permutations." In addition to these backcross methods, there is the double backcross (Walkof 1955, 1961), the inbred backcross (Wehrhan and Allard 1965; Dudley 1982; Cox 1984), and the congruity or alternating interspecific backcross (Haghighi and Ascher 1988; Barker et al. 1989; Superak et al. 1993). The double backcross is used to break linkage between two negatively correlated traits. The inbred backcross was initially designed to count genes, but it is now used to introgress needed genes from unadapted germplasm and improve adaptation in wide crosses. The congruity or interspecific backcross develops genetic bridges, increases fertility among interspecific crosses, or transfers desired genes between species.

Pedigree Methods

Many of the varieties now in production have resistance to one or more viruses and could be intermated to produce breeding lines with a broad spectrum of viral resistance. Selection for multiple virus resistance in the F_2 is the quickest way to develop a broad spectrum of resistance. Large populations are generally required with this method because the expected frequency of the desired homozygous genotype is only ¼ for any single locus. As the number of required independent loci (l) increases, the frequency of desirable genotypes decreases by $(\frac{1}{4})^l$ (Equation 1). Cross protection and diverse symptoms also make selection in the F_2 inaccurate, but F_3 progeny tests should reveal any inaccuracies.

Single seed descent (Figure 8) is based on the principle of equal fecundity. Two parents are mated to produce an F_2 base population, but only one offspring is derived from each F_2 individual and carried forward to the F_3 generation. Likewise, a single progeny from each F_3 is contributed to the F_4, and so on. Thus, in all future generations of inbreeding, every individual traces back to a single and different F_2 progenitor (Brim 1966; Empig and Fehr 1971). This practice distinguishes three unique properties of single seed descent: constant allele frequency, constant population size, and changing genotype frequencies over generations. The proportion of homozygosity (H) increases as g (Equation 1) increases, whereas the proportion of heterozygosity decreases as $1-H$. The proportion of homozygosity for

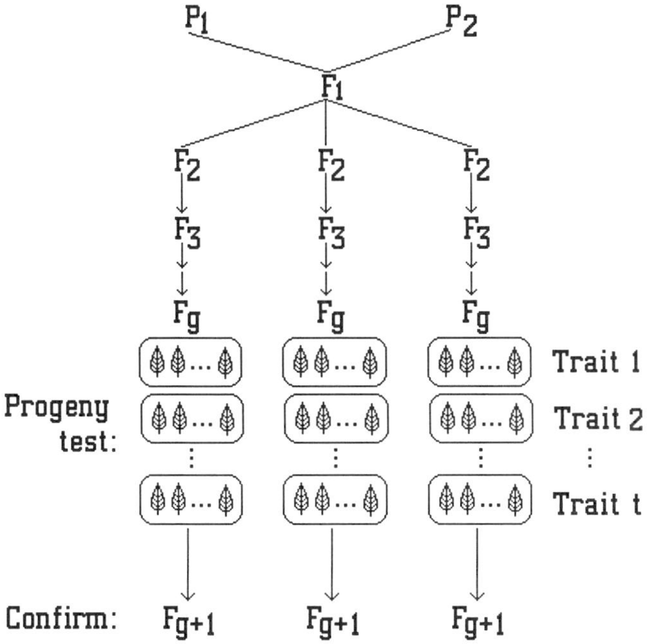

Figure 8. *A single seed descent program is used to incorporate a number of different traits. Selection, progeny testing, and confirmation are performed in the last two generations of the program, usually F_7 or F_8.*

all loci is 1.0 at F_∞. The original purpose of single seed descent was to develop inbred lines rapidly with a minimum of variation within lines and maximum variation among lines. For a quantitative trait, the probability of finding a transgressive segregate increases compared to the pedigree method, in which selection is intense in the early generations. In single seed descent, selection is usually practiced in F_7 or F_8 when "approximate homozygosity" is attained, although this is affected by the number of loci that condition a trait and their interaction (Snape and Riggs 1975; Hallauer and Miranda 1981).

In breeding for multiple viral resistance, F_2-derived F_g families are divided into subfamilies and each inoculated with a different virus (Figure 8). Resistance is then confirmed with a progeny test in F_{g+1} (Figure 8). Like the F_2 selection scheme, multiple viral resistance is identified in a single generation (F_g), with the work concentrated at the end of the program. Single seed descent is a slower process than

F_2 selection but requires a smaller population and produces genetically pure lines.

The single seed descent method can be modified so that multiple traits are selected separately in each generation (F_2, F_3, F_4, etc.) (Figure 9). In breeding for multiple viral resistance, the F_2 generation is inoculated with one virus and the resistant individuals carried forward to the F_3. In the F_3 generation, a second virus is inoculated and resistant individuals forwarded to the F_4. This process is repeated until all the desired traits are incorporated. At the end of the program, the number of lines to progeny test is much smaller than in either the F_2 or single seed descent schemes. The work is spread over all generations rather than concentrated in a single generation. This method is as slow as single seed descent, but there are fewer individuals to test with each succeeding generation. For the same probability of success, this method requires an initial F_2 population larger than single seed descent but smaller than F_2 selection.

The nested hierarchy (Cockerham 1954; Horner and Weber 1956; Wricke and Weber 1986) is a mating design originally developed to

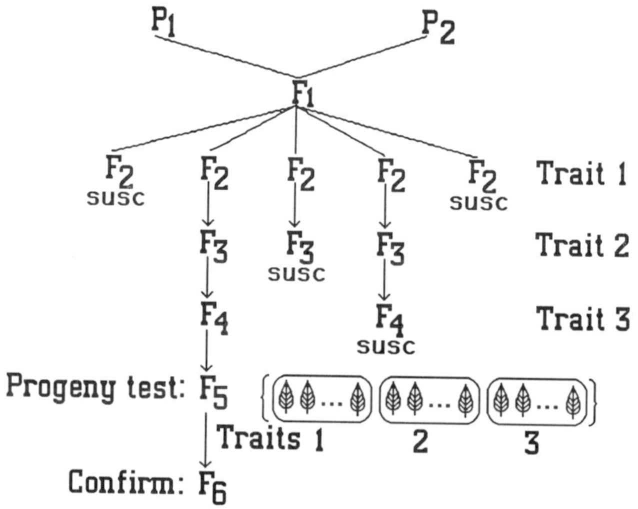

Figure 9. *In a modified single seed descent program, single traits are selected discretely in each generation. Progeny testing and confirmation of resistance are performed in the last two generations.*

partition genetic variances in self-pollinated crops but may be a useful breeding method for incorporating multiple virus resistance (Figure 10). It is similar to single seed descent and Goulden's (1939) modified pedigree method, and is perhaps best described as double, triple, or quadruple (etc.) seed descent. The number of divisions chosen is flexible, as is the number of times these divisions are made. This method differs from single seed descent in that the population grows geometrically larger as a function of the number and size of the divisions. Inbreeding proceeds at the same rate. Given the same objectives as a single seed descent program, the nested hierarchy should produce the same results. The size of the original F_2 population can be smaller than in single seed descent; however, it is absolutely critical that all needed genotypes or alleles be represented in

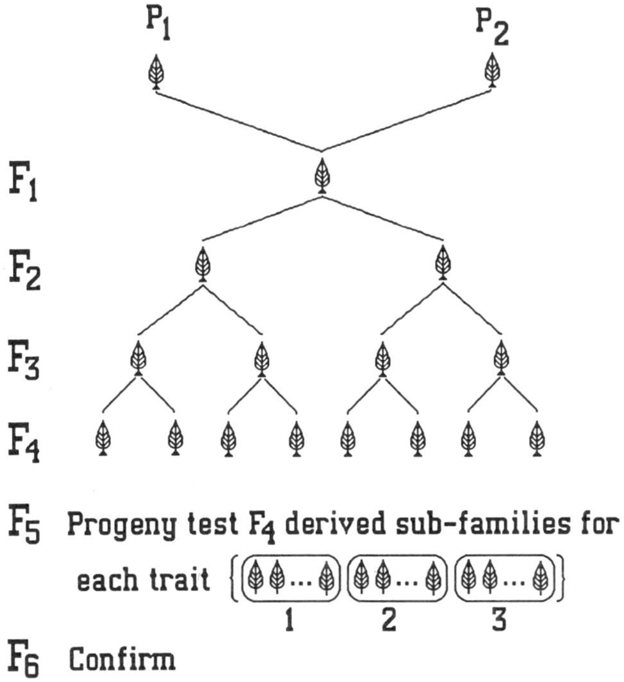

Figure 10. *The nested hierarchy is similar to single seed descent, but in this case two (or more) offspring are carried forward in each generation. Selection, progeny testing, and confirmation of the trait are determined in the last two generations.*

the F_2 generation. In the nested hierarchy, like single seed descent, multiple viral resistance is selected in the final generation and progeny is tested in the same way.

The nested hierarchy can be modified in a way analogous to the modified single seed descent. A single trait can be selected in each generation of inbreeding (Figure 11) until all resistant genotypes are identified. The size of the F_2 population will be larger than the nested hierarchy, but will not increase geometrically over generations because selection is applied in each generation. The number of lines tested in the final generation is smaller than in the single seed descent or nested hierarchy, given the same breeding objectives.

These pedigree and backcross methods are not presented at the exclusion of the bulk method, which is an easier and less expensive

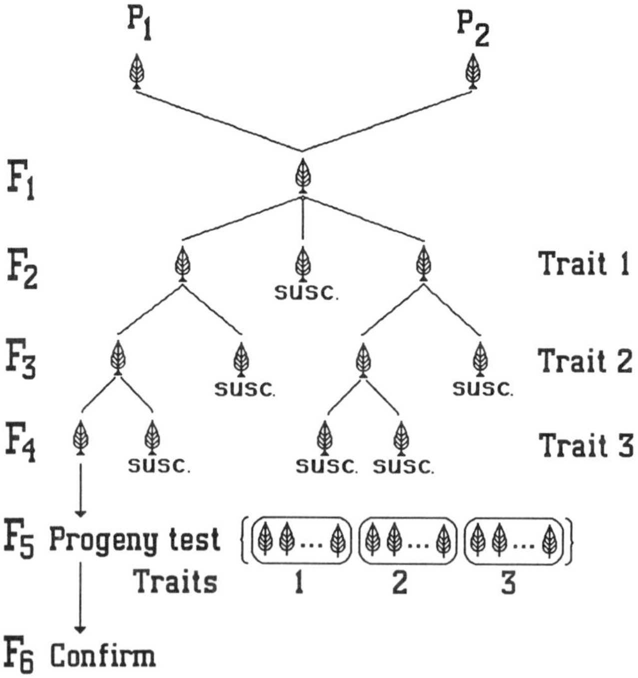

Figure 11. *In the modified nested hierarchy, single traits are selected discretely in each generation and the offspring from resistant individuals are carried forward. Progeny testing and confirmation of the trait are determined in the last two generations.*

way to deal with large populations. In environments in which viral diseases are pandemic and predictable year after year, bulk methods provide an opportunity to select for both viral resistance and environmental adaptation. These pedigree methods are intended to be flexible and serve as a guide for more creative breeding schemes that include selection for adaptation and other desirable traits. Breeding methods that fail to incorporate selection for adaptation and horticulturally important traits are likely to result only in the release of germplasm and breeding lines, rather than finished varieties.

Population Size

Binomial Distributions

The success of a particular selection technique and breeding method is a function of the population size, which should maximize the probability of success without excessive demands on time and resources. In this section we present a set of working equations that allow breeders to determine population size (n), and estimate the probability (P_α) that one, two, or more plants have the desired genotype (r) at an expected or constant genotypic frequency (f_d). These equations are drawn from binomial theory and are applicable to four selection situations. Selection for a single trait in a single generation provides the basic model and is used to construct three other multiple trait selection equations. Corollaries to the basic model include selection for different single traits separately in different generations; multiple traits jointly in a single generation; and subsets of two or more traits, each in different generations.

The use of binomial equations assumes that each trait is simply inherited, mutually exclusive, and qualitatively distributed with only two categories. These categories are based on either phenotypic, genotypic, or allelic frequency. In breeding for viral resistance in autogamous species, selection is ultimately practiced for the genotype, so genotypic frequency determines population size. Equations based on genotypic frequency are often more conservative than those based on either phenotypic or allelic frequency, and require larger populations. Conversely, population sizes determined by phenotypic or allelic frequency are commonly smaller, but can also include unwanted genotypes within the desired category.

Phenotypic and genotypic frequencies are equal when a trait is conditioned by additive gene action, but non-additive gene action skews phenotypic frequencies away from genotypic frequencies. For allelic frequencies, the probability of finding an individual with at least one copy of each desired allele is much higher than finding a unique genotype, particularly for polygenic traits (Sneep 1977). As the number of loci that condition a trait increases, differences in population sizes between allelic and genotypic based models become large.

The assumption of only two genotypic categories is valid for a single gene in a backcross generation, but inappropriate for F_2 populations segregating for mono/oligogenically inherited traits. These populations have three or more genotypes and follow multinomial distributions. However, these distributions are easily collapsed into two categories that include a single desired genotype in one category and all other genotypes in the second category. By collapsing populations into categories, differences in gene action, inheritance, penetrance, and other genetic phenomena are more easily manged. Application of binomial theory also requires that the inheritance of resistance be determined *a priori* to meet the assumption of known probability. This determination of expected genotype frequencies is essential for the assignment of genotypes to one category or another.

In cumulative binomial distributions, the probability of obtaining *r* plants of the desired genotype in a population of size *n* is defined by the expression (Mosteller et al. 1961; Larsen and Marx 1985)

$$P_\alpha = 1 - \sum_{x=n-r+1}^{n} \binom{n}{x} (f_d)^{n-x} (f_u)^x \qquad 2.$$

where P_α is the probability of success; n is the number of plants in the population; r is the number of individuals with the desired genotype, with $x=n-r+1$ as index of undesirable genotypes; f_d is the expected frequency of the desirable genotype; and f_u is the expected frequency of the undesirable genotype. Because the population is defined as a group of desirable and undesirable genotypes with constant frequency, then

$$f_u + f_d = 1. \qquad 3.$$

The binomial coefficients $\binom{n}{x}$ in Equation 2 are rewritten as

$$P_\alpha = 1 - \sum_{x=n-r+1}^{n} [n!/\,(r-1)!\,(n-r+1)!]\ (f_d)^{r-1}\ (f_u)^{n-r+1} \qquad 4.$$

with components defined above. For a single trait in a single generation, the probability of having *at least* one ($r \geq 1$) desired individual in a population of size n is defined by Snedecor and Cochran (1981) as

$$P_\alpha = 1-(f_u)^n, \qquad 5.$$

or more commonly as

$$P_\alpha = 1 - \{1 - (f_d)\}^n, \qquad 6.$$

as adapted from Equation 4. For the probability of *at least* two ($r \geq 2$) desired individuals:

$$P_\alpha = 1 - \{(f_u)^n + [n\ (f_d)\ (f_u)^{n-1}]\}; \qquad 7.$$

for at least three ($r \geq 3$) desired individuals:

$$P_\alpha = 1 - \{(f_u)^n + [n\ (f_d)\ (f_u)^{n-1}] + [((n_2-n)/2)\ (f_d)^2\ (f_u)^{n-2}]\}, \quad 8.$$

and so on, as Equation 4 is expanded. Because most breeders require a minimum of one individual with the desired genotype, Equations 5 and 6 are acceptable and easily solved for n. When $r \geq 2$, no closed-end solution exists for n in Equations 7 or 8. Values for n can be extracted from a summed binomial distribution table, but simpler tables that provide n given P_α and f_d are available (Table 1) (Harrington 1952; Sedcole 1977).

Sedcole (1977) also provides three computational methods to approximate population size when $r \geq 2$. The simplest method multiplies n for the probability of obtaining one of the desired genotype, by r. If 11 plants are needed in the F_2 for the probability of one individual, then 22 plants are required for $r \geq 2$; 33 for 3 etc. This approximation clearly overestimates population size (see Table 1). A fourth method that approximates n is defined for any given r (n'_r), as

$$n'_r = n_1 + [(r-1)n_1/2], \qquad 9.$$

where n_1 is the population size for a minimum of one desired individual as computed from Equation 5 or 6; and n'_r is the approximate value of n given r, with r defined previously. As an example, consider

Table 1. *The number of plants* (n) *needed in a population to recover a desired number of individuals* (r) *at a give genotypic frequency* (f_d) *and probability of success* (P_α)[a].

		r = THE NUMBER OF DESIRED INDIVIDUALS								
P_α	f_d	1	2	3	4	5	6	8	10	15
95%	1/2	5	8	11	13	16	18	23	28	40
	1/4	11	17	23	29	34	40	50	60	84
	1/8	23	37	49	60	71	82	103	123	172
	1/16	47	75	99	122	144	166	208	248	347
	1/32	95	150	200	246	291	334	418	500	697
	1/64	191	302	401	494	584	671	839	1002	1397
99%	1/2	7	11	14	17	19	22	27	32	45
	1/4	17	24	31	37	43	49	60	70	96
	1/8	35	51	64	77	89	101	124	146	198
	1/16	72	104	132	158	182	206	252	296	402
	1/32	146	210	266	318	368	416	508	597	809
	1/64	293	423	535	640	739	835	1020	1198	1623

[a]Adapted from Sedcole, J. R. 1977. Crop Sci. 17:667–668 with permission of the Crop Science Society of America.

a recessive gene in the F_2, with a $P_\alpha = 95\%$ and $f_d = 0.25$; the population size for $r \geq 1$ is 11 plants (Table 1). For $r = 2, 3, 4, 5, 6, 8, 10$, and 15, populations are approximated as $n' = 17, 22, 28, 33, 39, 49, 61$, and 84, respectively. This technique more closely approximates n than Sedcole's simplest procedure. In general, this method overestimates n at $P_\alpha = 99\%$, but underestimates n for $P_\alpha = 95\%$, at $r \geq 6$, and overestimates n as r becomes greater. These approximating methods are provided for frequencies (f_d) not covered in Table 1.

Single Trait: Single Generation

Selection for a monogenically inherited trait in a single generation is easily accomplished by a backcross method or selection in the F_2 generation. Consider an example with the alternate backcross and self procedure (Figure 2), in which a single recessive gene confers resistance to a particular virus. In the F_2 derived from each backcross generation, $f_d = 0.25$ and $f_u = 0.75$. Assuming the need for one

($r \geq 1$) homozygous recessive individual with a 95% (P_α) chance of success, the minimum population size (n) is computed with Equation 6, such that

$$0.95 = 1 - \{1 - (0.25)\}^n;$$

thus,

$$n = \ln(1-P_\alpha)/\ln(1-f_d) = \ln(0.05)/\ln(0.75) = 10.41 \simeq 11 \text{ plants.}$$

Eleven plants actually give a 95.78% chance of success. Based on the genotypic frequency (f_d), two or three plants of the 11 ($11 \cdot 0.25$) are expected to be homozygous recessive and resistant to the virus. The chance (P_α) of obtaining these two or three resistant individuals in a population of $n=11$ is 80% and 54.5%, respectively. If two or more ($r \geq 2$) resistant individuals are required at a $P_\alpha = 95\%$ then n is available from Table 1.

Multiple Traits: Single Generation

This basic probability equation can be extended to different selection problems. For two or more independent (i.e. unlinked) traits ($j=1$ to t) selected in a single generation, the probability that one or more individuals possesses these traits is the product of the f_{dj} ($f_{d1} \cdot f_{d2} \cdot \ldots \cdot f_{dt}$). For multiple viral resistance, the probability of obtaining a minimum of one individual with all desired traits is defined as

$$P_\alpha = 1 - \{1-(f_{d1} \cdot f_{d2} \cdot f_{d3} \cdot \ldots \cdot f_{dt})\}^n, \quad 10.$$

where P_α and n are defined above; and $f_{d1}, f_{d2}, f_{d3}, \ldots f_{dt}$ are the expected genotypic frequencies for each trait $j = 1 \ldots t$ in a single generation.

Thus,

$$P_\alpha = 1 - \{1-(\prod_{j=1}^{t} f_{dj})\}^n \quad 11.$$

This equation is appropriate for F_2 selection, single seed descent, and the parallel and multiple trait backcross procedures (Figures 6, 7, and 8). As an example, consider four viral diseases with resistance to each conditioned by a single dominant gene, and selection practiced in the F_2. The expected frequency of the homozygous dominant individuals is $(1/4)^4$ or 1/64. The probability that *at least* one individual has the desired genotype is

$$0.95 = 1 - \{1-(0.25 \cdot 0.25 \cdot 0.25 \cdot 0.25)\}^n;$$

thus,

$$n = \ln 0.05/\ln (1 - 1/64) = 190.2 \simeq 191 \text{ plants.}$$

In a population of 191 plants, an average of 3 individuals (191 · 1/64) are expected to have the desired genotype. With dominant gene action at all four loci, the heterozygous and homozygous individuals are indistinguishable in the F_2, and 60 plants ($0.75^4 \cdot 191$) are expected to have resistance. Cross protection and confounding symptoms additionally complicate the selection of homozygous resistant individuals. To avoid error, inoculation and selection are best practiced on four F_3 subfamilies derived from each of the 191 F_2 individuals.

For the nested hierarchy (Figure 10), Equation 11 is adjusted to reflect the number and size of divisions made in the pedigree, such that

$$P_\alpha = 1 - \{1-(\prod_{j=1}^{t} f_{dj})\}^{n(a^s)} \qquad 12.$$

where P_α and f_{dj} are defined above; n is the number of plants required in the F_2 generation; a is the size of the division; and s is the number of generations over which these divisions are made.

The nested hierarchy is perhaps best applied when insufficient numbers of individuals are available from the F_2 generation. Rare genotypes or allelic combinations may not be represented if the F_2 population is too small, however Equation 12 assumes that all possible genotypes are represented. The probability of having all the needed alleles represented in a large F_3 population derived from a few F_2 individuals is small; therefore, the size of the F_2 generation is critical. Equation 12 is an approximating equation that becomes less accurate as the number (s) and size (a) of the divisions increase.

Single Traits: Multiple Generations

Another permutation of the binomial equation is used for selecting an array of single traits, each in a different generation ($i = 1$ to g).

This selection technique is practiced in the modified single seed descent program (Figure 9), and is computationally very similar to selection for multiple traits in a single generation, such that

$$P_\alpha = 1 - \{1 - (\prod_{i=1}^{g} f_{di})\}^n \qquad 13.$$

where f_{di} is the frequency of the desired genotype in the ith generation, $i = 1 \ldots g$, where a given trait is selected; and all other components are defined above.

Consider a modified single seed descent program where three recessive genes (*aa*, *bb*, and *cc*) each confer resistance to viruses A, B, and C, respectively. If they are selected separately in the F_2, F_3, and F_4 generations, the desired genotypes occur at frequencies of 1/4, 3/8, and 7/16, respectively, in each generation. The initial number of F_2 plants required for at least one individual with the *aabbcc* genotype is computed as

$$P_\alpha = 1 - \{1 - (1/4 \cdot 3/8 \cdot 7/16)\}^n.$$

When P_α is set at 0.99, then

$$n = \ln(.01) / \ln(1 - 21/512) = 109.9 \simeq 110 \text{ plants.}$$

With a population of 110 plants, the general expectation is that 4 to 5 (110 · 21/512) individuals will have the desired genotype in F_4. If resistance were conferred by three dominant genes, heterozygous individuals would be carried forward into the next generation and a quarter of these would segregate into the homozygous resistant category. By the final generation, this would increase the number of resistant individuals in the population but would also increase the amount of progeny testing required. For the modified nested hierarchy (Figure 11), the problems and the computational adjustments are the same as those used for the nested hierarchy (Equation 12).

As an example, Scully et al. (1988) used the modified nested hierarchy to develop multiple viral resistance breeding lines of common beans (*Phaseolus vulgaris* L.) from an F_2 population of 1400 plants. A single division (s=1) of size 2 (a=2) was made in the F_3 generation. Selection was practiced for resistance to six viruses in generations F_2 through F_7. Single genes conditioned resistance to five of the viruses and one virus required two recessive genes, which was selected in the F_3. The F_2 population size was approximated as follows:

$$P_\alpha = 1-\{1-(1/4 \cdot 9/64 \cdot 7/16 \cdot 15/32 \cdot 31/64 \cdot 63/128)\}^{n(2^1)}$$

For an approximate P_α of 99.25%, 1422 plants were required in the F_2 generation.

Multiple Traits: Multiple Generations

The third permutation is best suited for a modified single seed descent program (Figure 9). It involves selection for subsets of two or more traits, with each subset in a different generation. In this case the size of the original F_2 population is defined as

$$P_\alpha = 1 - \{1 - (f_{d1(1)} \cdot f_{d2(1)} \cdot f_{dt(1)} \cdot \ldots \cdot f_{dt(g)})\}^n \qquad 14.$$

Thus,

$$P_\alpha = 1 - \{1 - (\prod_{i=1}^{g} \prod_{j(i)=1}^{t(g)} f_{dj(i)})\}^n \qquad 15.$$

where P_α and n are defined above; and $fd(j)i$ is the expected frequency of the jth trait nested within the ith generation. If selection is practiced for multiple traits, it is best to combine traits that do not compound selection for each other.

Conclusion

The concepts presented here are intended to provide a useful guide for the development of autogamous genotypes with multiple viral resistance. Consideration must be given to the inheritance of resistance, the statistical properties that influence genetic gain from selection, and the biology of the virus and its interaction with the host plant. The use of simply inherited broad spectrum virus resistance genes can further expedite the process of developing multiple viral resistant genotypes (Kyle and Provvidenti 1993). In addition to the standard breeding schemes, the multiple trait backcross, modified single seed descent, and both forms of the nested hierarchy were presented as supplemental breeding methods. They were developed as logical extensions of existing backcross and pedigree methods, and diversify the approaches for breeding for multiple virus resistance. A set of equations based on binomial theory were provided to determine population sizes for different breeding schemes. These equations can improve the efficiency of selection without overextending the time and resources allocated to a specific breeding objective.

Literature Cited

Allard, R. W. 1960. Principles of Plant Breeding. New York: Wiley and Sons Inc.

Baker, R. J. 1986. Selection Indices in Plant Breeding. Boca Raton, FL: CRC Press Inc.

Barker, T., G. Varughese, and R. Metzger. 1989. Alternative backcross methods for introgression of variability into triticale via interspecific hybrids. Crop Sci. 29:963–965.

Brim, C. A. 1966. A modified pedigree method of selection in soybeans. Crop Sci. 6:200.

Browning, J. A. 1980. Genetic protective mechanisms of plant-pathogen populations: Their coevolution and use in breeding for resistance. In Biology and Breeding for Resistance to Arthropods and Pathogens in Agricultural Plants. Ed. M. K. Harris. pp. 52–76. College Station, TX: Texas A&M University Press.

Cockerham, C. C. 1954. An extension of the concept of partitioning heredity variances for analysis of covariance among relatives when epistasis is present. Genetics 39:859–878.

Cox, T. S. 1984. Expectations of means and genetic variances in backcross populations. Theor. Appl. Genet. 68:35–41.

Day, P. R. 1974. Genetics of Host Parasite Interactions. San Francisco: W. H. Freeman Inc.

Dudley, J. W. 1982. Theory for the transfer of alleles. Crop Sci. 22:631–637.

Empig, L. T., and W. R. Fehr. 1971. Evaluation of methods for generation advance in bulk hybrid soybean populations. Crop Sci. 11:51.

Goulden, C. H. 1939. Problems in plant selection. In Proc. 7th Int. Genet. Congr. Ed. R. C. Punnet. pp.132–133. Cambridge, England: Cambridge University Press.

Gracen, V. E. 1982. Role of genetics in etiological phytopathology. Ann. Rev. Phytopathol. 20:219–233.

Haghighi, K. R., and P. D. Ascher. 1988. Fertile, intermediate hybrids between *Phaseolus vulgaris* and *P. acutifolius* for congruity backcrossing. Sexual Plant Reproduction 1:51–58.

Hallauer, A. R., and J. B. Miranda. 1981. Quantitative Genetics in Maize Breeding. Ames, IA: Iowa State University Press.

Harrington, J. B. 1952. Cereal breeding procedures. F.A.O. Development Paper No. 28. Rome: U.N. Food and Agricultural Organization.

Horner, T. W. and C. R. Weber. 1956. Theoretical and experimental study of self-fertilized populations. Biometrics 12:404–414.

Khush, G. S. 1980. Breeding for multiple disease and insect resistance in rice. In Biology and Breeding for Resistance to Anthropods and Pathogens in Agricultural Plants. Ed. M. K. Harris. pp. 341–345. College Station, TX: Texas A&M University Press.

Kyle, M. M., and R. Provvidenti. 1993. Genetics of multiple virus resistance in bean and pea. In Resistance to Viral Diseases of Vegetables: Genetics and Breeding. Ed. M. M. Kyle. Portland, OR: Timber Press.

Larsen, R. J. and M. L. Marx. 1985. An Introduction to Probability and Its Applications. Englewood Cliffs, NJ: Prentice-Hall, Inc.

Mayo, O. 1987. Theory of Plant Breeding. Oxford, England: Clarendon Press.

Mosteller, F., R. E. K. Rourke, and G. B. Thomas. 1961. Probability with Statistical Applications. London: Addison-Wesley Publ. Co., Inc.

Munger, H. M. 1993. Breeding for Viral Disease Resistance in Cucurbits. In Resistance to Viral Diseases of Vegetables: Genetics and Breeding. Ed. M. M. Kyle. Portland, OR: Timber Press.

Nelson, R. R. 1973. The use of resistance genes to curb population shifts in plant pathogens. In Breeding Plants for Disease Resistance: Concepts and Applications. Ed. R. R. Nelson. pp. 49–67. University Park, PA: Penn. State University Press.

Provvidenti, R. 1985. Lectures on the Resistance to Viral Diseases in Vegetables. International seminar on virus diseases of horticultural crops in the tropics. Council on Agriculture of the Republic of China.

Russell, G. E. 1978. Plant Breeding for Pest and Disease Resistance. London: Butterworth & Co. Ltd.

Sedcole, J. R. 1977. Number of plants necessary to recover a trait. Crop Sci. 17:667–668.

Scully, B. T., D. H. Wallace, and R. Provvidenti. 1988. Breeding common beans for multiple virus resistance. HortScience 23:115.

Sheridan, W. F., and J. K. Clark. 1987. Allelism testing by double pollination of lethal maize *dek* mutants. J. Heredity 78:49–50.

Simmonds, N. W. 1979. Principles of Crop Improvement. New York: Longman Group Ltd.

Smith, H. F. 1936. A discriminant function for plant selection. Ann. Eugen. 7:240–250.

Snape, J. W., and T. J. Riggs. 1975. Genetical consequences of single seed descent in the breeding of self-pollinated crops. Heredity 35:211-219.

Snedecor, G. W., and W. G. Cochran. 1981. Statistical Methods. 7th ed. Ames, IA: Iowa State University Press.

Sneep, J. 1977. Selection for yield in early generations of self-fertilizing crops. Euphytica 26:27–30.

Superak, T., B. T. Scully, M. M. Kyle, and H. M. Munger. 1993. Interspecific transfer of plant viral resistance. In Resistance to Viral Diseases of Vegetables: Genetics and Breeding. Ed. M. M. Kyle. Portland OR: Timber Press.

Suzuki, D. T., A. J. F. Griffiths, J. H. Miller, and R. C. Lewontin. 1985. An Introduction to Genetic Analysis. 3rd ed. New York: W. H. Freeman and Co.

Turner, H. N., and S. S. Y. Young. 1967. Quantitative Genetics in Sheep Breeding. Ithaca, NY: Cornell University Press.

Van Vleck, L. D. 1979. Notes on the Theory and Application of Selection Principles for Genetic Improvement of Animals. Dept. of Animal Science, Cornell University, Ithaca, NY.

Walkoff, C. 1955. An application of the double backcross method to tomato improvement. 14th Rept. Intern. Hort. Congr., Netherlands, pp. 252–259.

Walkoff, C. 1961. Improvement of tomato fruit size and maturity by backcross breeding. Can. J. Plant Sci. 41:24–30.

Wehrhan, C. and R. W. Allard. 1965. The detection and measurement of the effects of individual genes involved in the inheritance of a quantitative character in wheat. Genetics 51:109–119.

Wricke, G., and W. E. Weber. 1986. Quantitative Genetics and Selection in Plant Breeding. New York: Walter de Gruyter.

CHAPTER 9

Genetic Resistance that Reduces Disease Severity and Disease Incidence

STEWART M. GRAY & JAMES W. MOYER

Introduction

Resistance is the most effective, and in many cases the only, strategy for the control of diseases of plants caused by viruses. Heritable sources of resistance to one or more plant viruses have been identified in most major crop plants; however, the level of disease control and the type of resistance vary considerably. The foremost reason to identify or develop resistance is disease control. Nonetheless, it is important to distinguish between types of resistance that reduce disease severity within a plant from those that reduce disease incidence in a population of plants, i.e. crop. The distinction is obvious, but one that is seldom made, and this lack of distinction contributes to the confusion over terminology used to define and describe resistance (Cooper and Jones 1983; Travantzis 1984). Both can be effective strategies for disease control, and the actual resistance mechanisms may not be mutually exclusive.

A majority of investigations of host resistance to plant viruses have focused on reducing disease severity, i.e. a single plant's ability to

Dr. Gray is in the US Department of Agriculture/Agricultural Research Service and Department of Plant Pathology, Cornell University, Ithaca, NY 14853. Dr. Moyer is in the Dept. of Plant Pathology, North Carolina State University, Raleigh, NC 27695.

resist viral infection or minimize the impact of infection on growth and yield. The most effective types of resistance render plants immune to infection by the virus or immune to inoculation by the natural virus vector; the latter property has been termed *klendusity* (Jones and Cooper 1984). Although examples of immunity and klendusity exist, they represent extreme forms of resistance and have not been identified for most virus-host combinations. Fortunately, host plant resistance to plant viruses or their vectors may be expressed in many qualitative forms, each at various quantitative levels (Ponz and Bruening 1986; Fraser 1987; Jones 1987). Due to methods commonly used to select resistant genotypes, however, subtle forms of resistance are easily and often overlooked, even though they may effectively reduce disease severity and/or disease incidence.

For the purpose of this chapter, the term *resistance* is used as defined by Cooper and Jones 1983, i.e. there is an active response, presumably by the plant, that imposes constraints on the infectibility, multiplication, or movement of the virus. *Tolerance*, on the other hand, is a lack of disease response by a virus-infected host and does not necessarily constitute an active response by the host against the virus. We recognize that the terms *resistant* and *tolerant* are often used interchangeably and that there is not a consensus agreement of their definitions as they pertain to plant viral diseases.

Resistance as a Means to Reduce Disease Severity

The initial identification and selection for viral resistance in whole plants relies, more often than not, on a lack of or a reduction in expression of disease symptoms. The validity of using expression of disease symptoms as a selection criteria for viral resistance is determined in part by which plant tissues are affected and by which portion of the plant is marketed and for what use. Obviously the expression of symptoms on fruit would reduce the value and marketability of the fruit. In these cases, symptom expression is an important and useful screening technique. As the sole criteria for selection of resistant genotypes, however, it has many inherent problems. For example, symptom expression or severity is not always correlated with the ability of a virus to infect or multiply in a plant, nor is it always an indication of the reduction in plant growth or yield. The cowpea variety 'Iron', when infected with cowpea chlorotic mottle virus, expresses mild symptoms, yet the plants contain high amounts of virus and yield is significantly reduced. In contrast, the cowpea variety 'California Blackeye' expresses severe symptoms, but the plants contain little virus and there are no yield effects (Kuhn et al. 1981).

Clearly, the mechanisms of viral symptom expression are not well understood (Collmer and Howell 1993), and there is often no consistent relationship between symptom severity and the agronomic impact of the disease (Kuhn et al. 1981; Culver et al. 1987; Brown et al. 1987). Viral genes and gene products are involved, but time of infection, environment, and host plant also influence the expression of visual symptoms (Thompson and Hebert 1970; Paguio et al. 1987, 1988). Therefore, a reliance upon symptom expression as a measure of resistance without including other, perhaps more appropriate, phenotypic characteristics of disease development and viral pathogenesis often results in overlooking incomplete, but potentially useful forms of resistance.

In recent years, advances in technology have allowed for quantitative investigations of the mechanisms of host plant resistance to viral diseases. Studies of plant genotypes originally selected by the reduction or absence of symptoms have identified several distinct resistance mechanisms that limit virus invasion, replication, or translocation, thereby resulting in an overall reduction in virus titer and/or distribution (Ponz and Bruening 1986). As a result, useful techniques to rapidly and reliably screen for resistance have been developed that do not depend entirely on symptom expression (Barker and Harrison 1986; Stoddard et al. 1987; Seifers and Martin 1988; Kuhn et al. 1989; Dufour et al. 1989). Unfortunately, a majority of putative resistant genotypes continue to be selected on the basis of symptom expression, with little or no emphasis on the mechanisms of resistance.

Resistance as a Means to Reduce Disease Incidence

Investigations of resistance mechanisms at the organismal (plant) or cellular and molecular levels are essential; however, the benefits or consequences of various resistance mechanisms on disease epidemiology must not be neglected. As previously stated, host plant resistance can be used to control disease by reducing disease incidence in a population of plants. Studies on plant populations can examine the impact of various types of host plant resistance on disease severity and disease incidence, as well as determine the interactive effects of disease severity and incidence on crop growth and yield. In addition, resistance mechanisms effective at controlling viral disease at the population level are not limited to those affecting virus-host interactions. As most plant viruses depend upon specific vectors for interplant dispersal, the diversity of resistance mechanisms available to control viral disease incidence also includes those that disrupt vector-host or virus-vector interactions. Mechanisms of host plant resistance affecting viruses or their vectors may indirectly contribute to a significant reduction in the spread of the virus by lowering the efficiency by which the virus is transmitted among potential host plants (Zitter 1975; Lecoq et al. 1979; Cohen 1982; Romanow et al. 1986; Jones 1987; Berlinger and Dahan 1987). The resistance becomes, in effect, a resistance to virus transmission. When fewer plants become infected or there is a delay in the epidemic, the result is a lesser impact on crop yield. Host plant resistance to transmission can result in significant control of the incidence and spread of viruses (Lecoq and Pitrat 1983; Gray et al. 1986; Gunasinghe 1988; Lapoint et al. 1987; Habino et al. 1988), but only in recent years has it been recognized as a worthwhile mechanism for disease control.

In addition, studies on the epidemiological effects of host plant resistance may reveal potential shortcomings of putative resistant genotypes selected by lack of symptom expression. For example, tolerant plants (as defined by Cooper and Jones 1983) are often identified as resistant due to a lack of disease reaction by the plant. Tolerance can be extremely useful in terms of production and yield, but can be extremely detrimental in terms of disease management. The infected plants can serve as efficient sources of virus inoculum and viruliferous vector populations. Although the tolerant crop is protected from effects of the disease, nearby crops susceptible to the virus are exposed to a greater threat of disease.

Effective utilization of available forms of resistance that are expressed at various quantitative levels is facilitated not only by a basic understanding of the mechanisms and the target of the resistance within a single plant, but also by an understanding of how the resistance affects the epidemiology of the disease in a population of resistant plants. The remainder of this chapter summarizes results from a program developed to investigate resistance in muskmelon, *Cucumis melo*, to watermelon mosaic virus (WMV, formerly WMV-2). Based solely on symptomatology, the resistance may have been overlooked; however, investigations into the mechanisms of resistance and the epidemiological effects of the resistance, summarized here, have identified a valuable source of resistance that has been incorporated into a breeding program (H. M. Munger, personal communication). The multiple types of resistance identified and characterized reduced both disease severity within plants and disease incidence levels in the crop.

Multiple Forms of Resistance in *Cucumis melo* to Watermelon Mosaic Virus

WMV is a member of the potyvirus group that causes mosaic and mottle diseases of cucurbits and pea and infects numerous other hosts, including various leguminous, malvaceous, and chenopodiaceous weeds, ornamentals, and crop plants. The virus is naturally transmitted in a nonpersistent manner by at least 38 species of aphids (Purcifull et al. 1984). Watermelon mosaic disease occurs worldwide and causes significant losses in *Cucurbita pepo* L. and *C. melo* (Munger 1993). Resistance has not been reported in *C. pepo* (Provvidenti 1993), and our investigations summarized here were the first reports of resistance in *C. melo* to WMV. In addition, these studies are among the first to identify and characterize multiple forms of resistance to a virus and its vectors and to quantify the epidemiological effects of the various types of resistance.

"Suppressive" Virus Resistance in *Cucumis melo*

Suppressive virus resistance was the term used by Moyer et al. (1985) to define the response of the *C. melo* accession 91213 to infection by WMV. Accession 91213 is an eighth-generation inbred derived from PI 371795 (Kishaba et al. 1971). Moyer et al. (1985) determined that there was a reduction or a suppression in the amount of both WMV capsid protein and infectious virus accumulating in resistant 91213 plants relative to amounts in plants of the susceptible varieties 'Top Mark' and 'Hales's Best'. Although a positive correlation was found between increasing leaf age and increasing amount of capsid protein and infectious virus in WMV-susceptible plants, no such correlation was evident in the resistant 91213 plants. In addition, the decrease in infectious virus was greater than what would have been predicted from the decrease in levels of capsid protein, thus suggesting that assembly of infectious virions was affected or the virions were unstable.

Differences in the disease symptoms were also observed in the 91213 plants. Symptoms induced by WMV in plants of susceptible varieties (e.g. 'Hale's Best' or 'Top Mark') inoculated at the cotyledon or first leaf stage were initially apparent as a veinal chlorosis that evolved into chlorotic mottling and eventually general chlorosis. Symptoms in resistant 91213 plants inoculated at the cotyledon or first-leaf stage began as distinct chlorotic spots that expanded slowly. These chlorotic lesions did not always develop on the inoculated leaves, but they appeared on systemically infected leaves. On the fourteenth to seventeenth leaf on the inoculated runner, symptoms consistently changed to a distinct mosaic, but as these leaves matured, mild veinal chlorosis and diffuse chlorotic spotting or mild chlorotic mottling developed. Symptoms occasionally disappeared as the plant matured, and often runners from an infected plant would not develop viral symptoms (Moyer et al. 1985; Gray et al. 1988).

The distinct chlorotic areas and the mosaic pattern of symptom expression in 91213 plants were suggestive of a localization of the virus within the affected leaf tissue. WMV capsid protein could be detected only in the chlorotic areas of the leaf or from the light green areas of leaves showing mosaic symptoms, but not from surrounding unaffected or dark green tissue (Gray et al. 1988). Investigations on other viruses causing mosaic symptoms have shown high levels of virus and viral antigen in chlorotic areas and undetectable or low amounts in "dark green islands" (Atkinson and Matthews 1970; Loebenstein et al. 1977). Similar results were obtained from WMV-infected pumpkin (*C. pepo*), in which WMV induced mosaic and other symptoms in different leaves (Suzuki et al. 1989).

Localization of a virus is a common resistance mechanism in plants (Van Loon 1983; Fraser 1985), but the localization mechanism usually restricts the virus to the inoculated tissue before the virus can spread systemically. WMV was localized in the resistant 91213 leaves, but the ability of the virus to move systemically was not apparently affected. The cell-to-cell movement of plant viruses within leaves is distinct from the long-distance vascular transport between leaves (Atabekov and Dorokhov 1984; Hull 1989); therefore, whatever accounts for the resistance in 91213 plants appears to affect cell-to-cell movement but not vascular transport of WMV. A resistance mechanism affecting only one of two modes of virus transport in a plant is not unique to the WMV—*C. melo* system. Mechanisms of host resistance affecting vascular transport, but not cell-to-cell movement, have been described for at least two virus-host systems (Lei and Agrios 1986; Dufour et al. 1989; Law et al. 1989).

Subsequent studies were conducted to examine WMV pathogenesis in resistant and susceptible tissue and to further test the hypothesis that one mechanism of resistance in 91213 functions to limit cell-to-cell movement. Few studies have been made on viral pathogenesis in whole plants because the asynchronous infection of leaf cells by the virus makes identifying sequential events in the viral infection cycle difficult (Berna et al. 1986). An alternative to whole plants has been the use of protoplasts infected *in vitro*, although complete synchrony of protoplast infection is unlikely (Sugimura and Matthews 1981). The response of protoplasts from resistant plants to viral infection is often different from that of whole plants (Takebe 1977; Sarkar 1977), and protoplasts are not appropriate for investigating resistance mechanisms that interfere with cell-to-cell movement or vascular transport. A more appropriate experimental system is the near synchronous plant viral infection of whole plants, which has been achieved via a differential temperature treatment (Dawson and Schlegel 1973; Dawson et al. 1975; Dorokhov et al. 1981; Roberts and Wood 1981). This method was adapted for use in the *C. melo*–WMV system (Gray et al. 1988). Briefly, lower inoculated leaves were maintained at a temperature permissive for viral replication and systemic movement (25°C). Upper, uninoculated leaves were maintained at a low temperature (5°C) that allowed systemic movement of the virus into the tissue, but inhibited viral replication in the infected cells. Following a 3–7 day differential temperature treatment, the entire plant was transferred to the permissive temperature (25°C) and viral replication began and was synchronized in the upper, systemically infected leaves present on the plant during the differential temperature treatment.

Differential temperature manipulation of plants of the susceptible variety 'Top Mark' did not detectably alter the systemic infection. The type of disease symptom and the accumulation of capsid protein and infectious virus were similar in differential-temperature-treated plants and control plants grown at 25°C (Gray et al. 1988). In addition, the relationship between levels of infectious virus and capsid protein was similar to that reported by Moyer et al. (1985) for greenhouse-grown plants. In contrast, differential temperature manipulation of 91213 plants clearly altered the response to the virus. The symptom, discrete chlorotic lesions expressed in the treated leaves was similar to that of the untreated leaves, but the number of chlorotic lesions was significantly increased on the differential-temperature-treated leaves (Gray et al. 1988). Thus, differential temperature treatment of resistant 91213 plants resulted in a uniform distribution of lesions and an increase in the amount of symptomatic infected tissue (Gray et al. 1988). This allowed multiple small symptomatic tissue samples to be removed from one synchronously infected leaf and assayed for the accumulation of virus and viral-encoded proteins at multiple time points following infection. The symptomatic tissue, i.e. chlorotic lesions, on an infected 91213 leaf growing at 25°C was limited and developed at various times, thus prohibiting any meaningful time course study of viral protein accumulation within a leaf.

Initial studies in which the differential temperature treatment was used revealed that the levels and temporal pattern of accumulation of capsid protein from symptomatic treated leaves of 91213 and susceptible plants were not significantly different (Gray et al. 1988). This supported our hypothesis that the reduction of capsid protein in 91213 plants relative to 'Top Mark' plants, grown under normal conditions (Moyer et al. 1985; Gray et al. 1988), was due to fewer infected cells (i.e. a reduction in cell-to-cell spread of virus), rather than some interference with production of mature capsid protein. The levels of infectious virus were as much as fourfold lower in 91213 than in 'Top Mark' after the differential temperature treatment, however (S. M. Gray, unpublished data). These data further supported the hypothesis of Moyer et al. (1985) that the resistance operating in 91213 limited the production or stability of infectious virus in addition to limiting cell-to-cell movement of the virus.

The temporal pattern of accumulation of a second WMV-encoded protein was quantified concurrently from the same leaf tissue as the capsid protein. The cylindrical inclusion protein (CIP) aggregates in the cytoplasm of infected cells to form pinwheel or cytoplasmic inclusion bodies (CIs) characteristic of potyviruses and other plant viruses (Christie and Edwardson 1977; Martelli and Russo 1977). The actual function of the CIP or CIs has not been definitively determined, but several studies have associated CIs with plasmodesmata (Lawson et al. 1971; Andrews and Shalla 1974; Langenberg 1986; Lin et al. 1988). Langenberg (1986) has proposed that CIs assist in the cell-to-cell movement of viral particles through plasmodesmata. This theory was supported in part by a recent study on passion fruit woodiness virus. Lin et al. (1988) determined that CIP originates on the surface of the endoplasmic reticulum, and that as the inclusion bodies are formed they attach at right angles to the cell wall covering, in some cases, plasmodesmata. Virion aggregates were consistently associated with the CIs. Virions or CIs were not visualized in the plasmodesmata; however, subunits of both CIP and capsid protein were detected, thus suggesting that the two proteins move through the plasmodesmata in subunit form.

Preliminary studies on the accumulation of CIP in the resistant 91213 indicated that the expression or stability of the CIP was affected (S. M. Gray, unpublished data). CIP accumulated at similar rates and to similar levels during the first eight days in resistant and susceptible plants. Although CIP continued to increase in the susceptible plants, it declined precipitously in the resistant plants (Figure 1). An overall reduction in levels of CIP and/or in its biological activity may alter the ability of virus to move through plasmodesmata and thus restrict the viral infection to slowly expanding chlorotic lesions. Nonetheless, investigations on the accumulation of other nonstructural proteins and ultrastructural studies on the appearance and location of CIs in infected cells of the resistant and susceptible *C. melo* genotypes are needed. When these additional studies are completed, conclusions regarding the association between the reduced accumulation of CIP and the apparent reduction in cell-to-cell movement can be made.

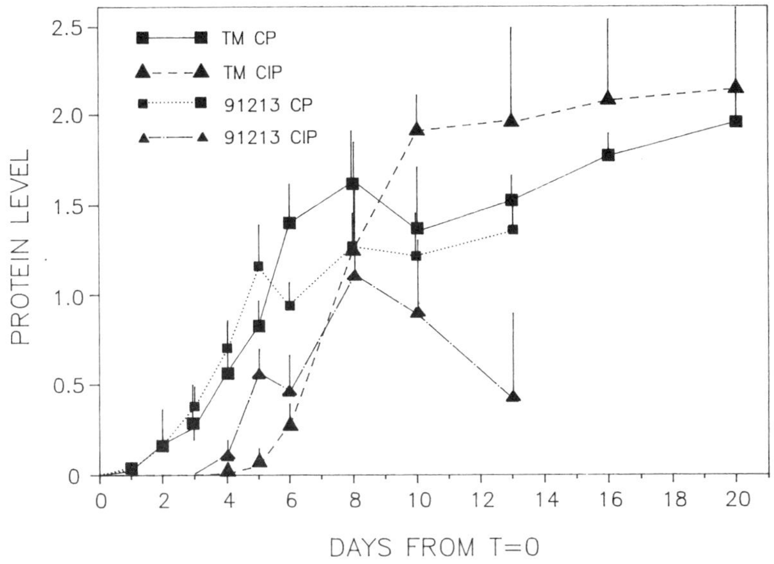

Figure 1. *Relative amounts of WMV capsid protein (CP) and cytoplasmic inclusion protein (CIP) quantified from differential temperature treated 'Top Mark' (TM) or resistant 91213 plants. T=0 refers to the time in days at which the inoculated plants were removed from the nonpermissive lower temperature and moved to 25°C. Protein concentrations were quantified from immunoblots by using a densitometer. Protein level is represented by the area under the single peak of the densitometer tracing.* Vertical bars *represent standard errors.*

Resistance to Aphid Vectors

A majority of plant viruses are dependent upon specific vectors for interplant dispersal; therefore, an alternative strategy to control the incidence of viral disease is to reduce the vector populations feeding or developing on viral host plants. One method to achieve this would be through host plant resistance to the vector. Vector resistance has been associated with a decreased incidence of viral infection on dozens of different vector-virus combinations (see Jones 1987). The actual mechanisms of resistance are not well defined in most cases; however, two broad categories have been defined (Smith 1989): (1) antixenosis, mechanisms of resistance that adversely affect vector behavior, (2) antibiosis, mechanisms of resistance that reduce the ability of the vector to survive and reproduce on the plant. Antibiosis and antixenosis are often expressed concurrently in a single plant, and their effects on vectors and virus transmission may not be mutually exclusive.

Resistance to the melon aphid, *Aphis gossypii* Glover, in the *C. melo* breeding line LJ90234 (inbred from PI 371795), was first reported by Kishaba et al. (1971) and shown to be heritable by Bohn et al. (1972). Peak populations of *A. gossypii* developing on field grown susceptible 'Top Mark' plants reached greater than 1000/leaf, whereas the average number developing on resistant LJ90234 plants was only 25.4/leaf (Kennedy et al. 1975). Subsequent studies to define the resistance were conducted by using the resistant breeding line 91213 also inbred from PI 371795 and 'Top Mark' (Kennedy and Kishaba 1976). Although populations of *A. gossypii* could be maintained on 91213 plants for several generations without the danger of extinction, aphid mortality during the reproductive period was higher and fecundity was lower relative to aphid populations developing on susceptible plants, thus indicating an antibiosis-type of resistance. In addition, fewer alates (winged morphs) were produced on resistant plants.

An antixenosis component to the aphid resistance in 91213 resulted in fewer aphids selecting the resistant line as a suitable host and in an increase in interplant movement (Kennedy and Kishaba 1977). Additional studies on the feeding behavior of aphids on 91213 plants revealed that a significantly greater percentage of aphid feeding probes led to stylet contact with the phloem cells, but a smaller proportion of the phloem contacts led to ingestion. In addition, the duration of periods of ingestion was two to three times shorter on resistant than on susceptible plants (Kennedy et al. 1978).

The effects of aphid resistance operating in 91213 on the epidemiology of plant viruses was not an objective of the previously described studies, but several inferences can be drawn. The alteration in feeding behavior is unlikely to have an effect on the transmission of nonpersistent viruses, e.g. WMV, since they are acquired and inoculated in a matter of seconds; however, the alteration in probing and feeding behavior is likely to be partially responsible for the reduction in aphid fecundity and the increase in aphid mortality. The reduction in aphid populations would contribute to a significant reduction in available aphid vectors overall, especially alate aphids. This would limit the spread of virus within the resistant crop as well as to other nearby susceptible crops. Additionally, since the resistance lowers the reproductive capacity of the aphids but does not result in their extinction, it is not likely to exert selection pressure for the development of aphid biotypes capable of overcoming the resistance. Also, lower population levels would minimize feeding damage. A potentially detrimental effect of the aphid resistance is the nonpreference of the aphids for the 91213 plants as a host. Viruliferous aphids may move more frequently among plants, potentially probing and inoculating each plant visited (Kennedy 1976).

Resistance to Viral Transmission—Epidemiological Consequences

The aphid resistance and the viral resistance discussed above were not genetically linked and have been separated. An aphid resistant 'Top Mark' variety ('AR-Top Mark') was developed by selecting for resistance to *A. gossypii* in a backcross breeding program in which 'Top Mark' was used as the recurrent parent and 91213 as the source of aphid resistance (A. N. Kisaba and G. W. Bohn, USDA/ARS, Riverside, CA). The existence of varieties carrying either aphid resistance alone or both aphid and viral resistance has allowed comparisons of the effects of the two types of resistance on WMV transmission by aphids (Romanow et al. 1986) and WMV epidemiology in a muskmelon crop (Gray et al. 1986).

The effects of aphid resistance alone versus aphid and viral resistance together on acquisition and inoculation of WMV were assessed by using two efficient vectors, *A. gossypii*, the only aphid vector that colonizes muskmelon, and *Myzus persicae* Sulzer, a noncolonizing aphid. As previously discussed, the viral resistance resulted in a limited distribution of virus-infected tissue and a further reduction in infectious virus within that tissue. This caused a significant reduction in the efficiency of transmission by reducing the acquisition efficiency of WMV by both aphid vectors (Romanow et al. 1986). Whether fewer aphids acquired a sufficient amount of virus from 91213 source leaves to make infection likely, or whether aphids acquired less virus, thereby lowering the probability of transmission for each individual, was not determined. The reduction in acquisition efficiency was a general form of resistance effective for both *C. melo* colonizing and noncolonizing (transient) aphid vector species. The viral resistance had no detectable effect on the inoculation efficiency of either aphid vector.

Resistance to aphids, expressed in both 91213 and 'AR-Top Mark', reduced the inoculation efficiency of *A. gossypii*, a colonizing aphid, but did not affect inoculation efficiency of noncolonizing aphid vectors. These results were similar to those of Lecoq et al. (1980) and Risser et al. (1981), who found that genes conferring resistance to *A. gossypii* in *C. melo* also selectively inhibited inoculation by *A. gossypii* of WMV, papaya ringspot virus, cucumber mosaic virus, and zucchini yellow mosaic virus. The mechanisms of the aphid resistance, expressed in both 91213 and 'AR-Top Mark', are unknown. Altering the probing behavior of *A. gossypii* (Kennedy et al. 1978) may interfere with the ability of *A. gossypii* to deposit virus in sites appropriate for initiation of viral infection.

The transmission of aphid-borne, nonpersistent viruses has been shown to be reduced in varieties or breeding lines of several hosts that possess some form of resistance to viral infection (Anzola et al. 1982; Thresh 1984; Barker and Harrison 1986) or to aphid vectors (Jones 1976; Lecoq et al. 1980). The epidemiological significance of these types of resistance has been described qualitatively for several host-virus-vector systems (Bawden and Kassanis 1946; Lecoq and Pitrat 1983), but there is little quantitative information available.

Our field study provided the first quantitative data on the spatial and temporal development of epidemics induced by WMV in a susceptible variety of *C. melo* ('Top Mark') and in genotypes possessing either aphid resistance ('AR-Top Mark') or aphid resistance coupled with suppressive viral resistance (Gray et al. 1986). Final virus incidence, assessed by ELISA in the aphid-resistant genotype and the aphid/virus resistant genotype, averaged 11% and 33% lower, respectively, than in the susceptible genotypes during a spring planting. The rate of disease progress was also significantly reduced in both resistant genotypes relative to the susceptible genotype. Infected plants of all three muskmelon genotypes were consistently observed to occur in a clustered pattern, but the clusters of plants were more loosely defined for the resistant genotypes.

The effectiveness of the resistance at reducing disease incidence was related to the seasonal phenology of aphid vector populations and the source of virus inoculum. In the spring, *A. gossypii* was the major vector, and a majority of the infections were the result of secondary movement of virus acquired from initial infection foci and transmitted to nearby plants. Limited sources of inoculum in surrounding areas and resistance to inoculation reduced the initial number of plants that become infected by viruliferous *A. gossypii* immigrating into the crop. The resistance to inoculation further reduced secondary spread of the virus within the crop. In addition, the antixenosis component, i.e. increased interplant movement, resulted in more loosely defined clusters of infected plants. Finally, less efficient acquisition from the reduced virus titer and limited viral distribution within infected plant tissue also contributed to a reduction in secondary spread of the virus by all aphid vectors.

In contrast, during summer planting, the final disease incidence and the rate of disease progress were not significantly different among genotypes and the infected plants occurred in a random spatial pattern. The ability of the aphid and viral resistance to be overcome can also be explained by the seasonal phenology of aphid vector populations and the source of virus inoculum. The number of aphids alighting on the summer crop increased significantly, and a majority were not *A. gossypii* and thus were unaffected by the aphid resistance, i.e. reduction in inoculation efficiency. The effectiveness of the viral resistance, which acts to reduce acquisition efficiency, depends in part on the availability and proximity of virus sources outside the resistant fields and the size and shape of the resistant fields. WMV is a limiting factor for squash and melon production during late summer and fall in the central region of North Carolina where these studies were conducted. There would have been a significant increase in the amount of WMV inoculum outside the test area, and a majority of the aphids alighting on the crop would have acquired virus from a source located outside the field of resistant plants. Large numbers of viruliferous, non-*A. gossypii*, vectors alighting in relatively small plots rapidly infected all of the plants through primary spread of the virus. The viral resistance, which reduces acquisition efficiency, is only effective if the epidemic is related to secondary virus spread and is of little epidemiological importance when a majority of the plants are infected due to primary spread.

Conclusions

These investigations have identified resistance in *C. melo* to WMV and have added to the general knowledge about the mechanisms of incomplete or partial host plant resistance. Several resistance mechanisms were identified in the *C. melo* line 91213. The mechanisms directly affect viral pathogenesis and disease development through both host-virus or host-vector interactions. Indirectly, the efficiency of virus transmission by aphid vectors is affected, thereby reducing the incidence of viral infection in plant populations. The direct effects of the various resistance mechanisms act to minimize the consequences of disease within single plants, thus resulting in better growth and yield. The indirect effects on transmission act to minimize disease impact at the population level, i.e. reduce the number of diseased plants, and slow the rate at which plants become diseased, thereby delaying the epidemic.

More importantly, these studies illustrate the need for understanding the complex virus-vector-plant interactions in order to effectively exploit various types of resistance. In addition, some basic information on the ecology of each virus-vector-plant system must be obtained prior to designing and implementing control strategies. The limitations of the individual resistance mechanisms must be explored before the benefits of host plant resistance in controlling the incidence and severity of viral disease can be assured.

The resistance to WMV identified in 91213 limits the distribution of virus infected tissue and reduces the overall virus titer. This and similar types of resistance in other virus-host systems can be beneficial in that they reduce disease severity in the plant and reduce the impact of the disease on plant growth and yield. In addition, these types of resistance reduce the acquisition efficiency of aphid vectors feeding on the resistant crop. Such resistance is general, in that it affects all aphid vectors and can effectively reduce disease incidence in the crop, but only if secondary spread of the virus from infected sources within the resistant field is of paramount importance. This type of resistance will have little benefit if a majority of the disease incidence is a result of viruliferous vectors moving into the crop.

The aphid resistance identified in the 91213 and 'AR-Top Mark' genotypes had no effect on disease severity in a plant, but it did act to reduce overall aphid populations. More importantly, it resulted in a reduction in the ability of *A. gossypii* to inoculate plants. Resistance to inoculation can reduce overall disease incidence in the protected crop, and extreme resistance to inoculation, i.e. klendusity, can effectively result in the crop's immunity to the viral disease if seedborne sources are eliminated. Although the resistance to WMV inoculation by *A. gossypii* was effective in a spring planting of muskmelon in North Carolina, it was effective only because *A. gossypii* was the major vector. The limitation to this type of resistance is its vector specificity and the requirement of previous knowledge of seasonal phenology of virus vectors expected in the area where the crop is to be grown. The benefits are its effectiveness against both primary and secondary spread of the virus as well as the reduction in damage to the crop by the vector alone.

The lack of effective, economical measures to control most plant virus diseases emphasizes the extreme urgency to identify and develop effective virus-resistant germplasm. As our understanding of virus-vector-host interactions increases, so does the potential to develop novel strategies to interfere with these processes. The use of incomplete host plant resistance to reduce the incidence of plant viral diseases has gained acceptance and has been successfully applied in many crops. Many problems remain to be overcome, however, most notably in the detection and exploitation of resistance mechanisms. Many resistance mechanisms are discovered serendipitously during a search for other agronomic traits. Small scale screening methods dependent upon symptom expression may not detect subtle forms of resistance, or these forms may not be recognized by those who do not have the resources to test for them. The techniques are available to identify and exploit all types of resistance mechanisms, but the degree of effectiveness will be correlated with the degree of interdisciplinary effort given toward understanding the complex virus-vector-host relationships that drive virus disease epidemics.

Literature Cited

Andrews, J. H., and T. A. Shalla. 1974. The origin, development, and conformation of amorphous inclusion body components in tobacco etch virus-infected cells. Phytopathology 64:1234–1243.

Anzola, D., C. P. Romaine, L. V. Gregory, and J. E. Ayers. 1982. Disease response of sweet corn hybrids derived from dent corn resistant to maize dwarf mosaic virus. Phytopathology 2:601–604.

Atabekov, J. G., and Yu. L. Dorokhov. 1984. Plant virus-specific transport function and resistance of plants to viruses. Adv. Virus Res. 29:313–364.

Atkinson, P. H., and R. E. F. Matthews. 1970. On the origin of dark green tissue in tobacco leaves infected with tobacco mosaic virus. Virology 40:344–356.

Barker, H., and B. D. Harrison. 1986. Restricted distribution of potato leafroll virus antigen in resistant potato genotypes and its effect on transmission of the virus by aphids. Ann. Appl. Biol. 109:595–604.

Bawden, F. C., and B. Kassanis. 1946. Varietal differences in susceptibility to potato virus Y. Ann. Appl. Biol. 33:46–53.

Berlinger, M. J., and R. Dahan. 1987. Breeding for resistance to virus transmission by whiteflies in tomatoes. Insect Sci. Applic. 8:783–784.

Berna, J., J. P. Briand, C. Stussi-Garaud, and T. Godefroy-Colburn. 1986. Kinetics of accumulation of the three non-structural proteins of alfalfa mosaic virus in tobacco plants. J. Gen. Virol. 67:1135–1147.

Bohn, G. W., A. N. Kishaba, and H. H. Toba. 1972. Mechanisms of resistance to melon aphid in a muskmelon line. HortScience 7:281–282.

Brown, K. J., J. D. Mihail, and M. R. Nelson. 1987. Effects of cotton leaf crumple virus on cotton inoculated at different growth stages. Plant Dis. 71:699–703.

Christie, R. G., and J. R. Edwardson. 1977. Light and electron microscopy of plant virus inclusions. Florida Agr. Exp. Station Monogr. Ser. 9.

Cohen, S. 1982. Resistance to transmission of aphid-borne nonpersistent viruses in vegetables. Acta Hort. 127:117–124.

Collmer, C. W., and Howell, S. H. 1993. Mechanisms of plant viral pathogenesis. In Resistance to Viral Diseases of Vegetables: Genetics and Breeding. Ed. M. M. Kyle. Portland, OR: Timber Press.

Cooper, J. I., and A. T. Jones. 1983. Responses of plants to viruses: Proposals for the use of terms. Phytopathology 73:127–128.

Culver, J. N., J. L. Sherwood, and H. A. Melouk. 1987. Resistance to peanut stripe virus in *Arachis* germplasm. Plant Dis. 71:1080–1082.

Dawson, W. O., and D. E. Schlegel. 1973. Differential temperature treatment of plants greatly enhances multiplication rates. Virology 53:476–478.

Dawson, W. O., D. E. Schlegel, and M. C. Y. Lung. 1975. Synthesis of tobacco mosaic virus in intact tobacco leaves systemically inoculated by differential temperature treatment. Virology 65:565–573.

Dorokhov, Y. L., N. A. Miroshnichenko, N. M. Alexandrova, and J. G. Atabekov. 1981. Development of systemic TMV infection in upper noninoculated tobacco leaves after differential temperature treatment. Virology 108:507–509.

Dufour, O., A. Palloix, K. G. Selassie, E. Pouchard, and G. Marchoux. 1989. The distribution of cucumber mosaic virus in resistant and susceptible plants of pepper. Can. J. Bot. 67:655–660.

Fraser, R. S. S. 1985. Mechanisms involved in genetically controlled resistance and virulence: Virus diseases. In Mechanisms of Resistance to Plant Diseases. Ed. R. S. S. Fraser. Lancaster: Nijhoff and Junk.

Fraser, R. S. S. 1987. Genetics of plant resistance to viruses. In Plant Resistance to Viruses, Ciba Found. Symp. 133:6–22. London: J. Wiley and Sons.

Gray, S. M., J. W. Moyer, and G. G. Kennedy. 1988. Resistance in *Cucumis melo* to watermelon mosaic virus 2 correlated with reduced virus movement within leaves. Phytopathology 78:1043–1047.

Gray, S. M., J. W. Moyer, G. G. Kennedy, and C. L. Campbell. 1986. Virus-suppression and aphid resistance effects on spatial and temporal spread of watermelon mosaic virus 2. Phytopathology 76:1254–1259.

Gunasinghe, U. B., M. E. Irwin, and G. E. Kampmeier. 1988. Soybean leaf pubescence affects aphid vector transmission and field spread of soybean mosaic virus. Ann. Appl. Biol. 112:259–272.

Habino, H., R. D. Daquioag, P. Q. Cabauatan, and G. Dahal. 1988. Resistance to rice tungro spherical virus in rice. Plant Dis. 72:843–847.

Hull, R. 1989. The movement of viruses in plants. Ann. Rev. Phytopathol. 27:213–240.

Jones, A. T. 1976. The effect of resistance to *Amphorophora rubi* in raspberry (*Rubus ideaus*) on the spread of aphid-borne viruses. Ann. Appl. Biol. 82:503–510.

Jones, A. T. 1987. Control of virus infection in crop plants through vector resistance: A review of achievements, prospects and problems. Ann. Appl. Biol. 111:745–772.

Jones, A. T., and J. I. Cooper. 1984. A reply to the comments of S. M. Tavantzis. Phytopathology 74:381.

Kennedy, G. G. 1976. Host plant resistance and the spread of plant viruses. Environ. Entomol. 5:827–832.

Kennedy, G. G., and A. N. Kishaba. 1976. Bionomics of *Aphis gossypii* on resistant and susceptible cantaloupe. Environ. Entomol. 5:357–361.

Kennedy, G. G., and A. N. Kishaba. 1977. Response of alate melon aphids to resistant and susceptible muskmelon lines. J. Econ. Entomol. 70:407–410.

Kennedy, G. G., A. N. Kishaba, and G. W. Bohn. 1975. Response of several pest species to *Cucumis melo* L. lines resistant to *Aphis gossypii* Glover. Environ. Entomol. 4:653–657.

Kennedy, G. G., D. L. McLean, and M. G. Kinsey. 1978. Probing behavior of *Aphis gossypii* on resistant and susceptible muskmelon. J. Econ. Entomol. 71:13–16.

Kishaba, A. N., G. W. Bohn, and H. H. Toba. 1971. Resistance to *Aphis gossypii* in muskmelon. J. Econ. Entomol. 64:935–947.

Kuhn, C. W., F. W. Nutter, and G. B. Padgett. 1989. Multiple levels of resistance to tobacco etch virus in pepper. Phytopathology 79:814–818.

Kuhn, C. W., S. D. Wyatt, and B. B. Brantley. 1981. Genetic control of symptoms, movement, and virus accumulation in cowpea plants infected with cowpea chlorotic mottle virus. Phytopathology 71:1310–1315.

Langenberg, W. G. 1986. Virus protein association with cylindrical inclusions of two viruses that infect wheat. J. Gen. Virol. 67:1161–1168.

Lapoint, S. L., W. M. Tingey, and T. A. Zitter. 1987. Potato virus Y transmission reduced in an aphid-resistant potato species. Phytopathology 77:819–822.

Law, M. D., J. W. Moyer, and G. A. Payne. 1989. Effect of host resistance on pathogenesis of maize dwarf mosaic virus. Phytopathology 79:757–761.

Lawson, R., S. S. Hearon, and F. F. Smith. 1971. Development of pinwheel inclusions associated with sweet potato russet crack virus. Virology 46:453–463.

Lecoq, H., S. Cohen, M. Pitrat, and G. Labonne. 1979. Resistance to cucumber mosaic virus transmission by aphids in *Cucumis melo*. Phytopathology 69:1223–1225.

Lecoq, H., G. Labonne, and M. Pitrat. 1980. Specificity of resistance to virus transmission by aphids in *Cucumis melo*. Ann. Phytopathol. 12:139–144.

Lecoq, H., and M. Pitrat. 1983. Field experiments of the integrated control of aphid-borne viruses in muskmelon. In Plant Virus Epidemiology. Ed. R. T. Plumb and J. M. Thresh. Oxford, UK: Blackwell Scientific Publications.

Lei, J. D., and G. N. Agrios. 1986. Mechanisms of resistance in corn to maize dwarf mosaic virus. Phytopathology 76:1034–1040.

Lin, N., N. Wang, and Y. Hsu. 1988. Sequential appearance of capsid protein and cylindrical inclusion protein in root-tip cells following infection with passion fruit woodiness virus. J. Ultrastruct. Res. 100:201–211.

Loebenstein, G., J. Cohen, S. Shabtai, R. H. A. Coutts, and R. K. Wood. 1977. Distribution of cucumber mosaic virus in systemically infected tobacco leaves. Virology 81:117–125.

Martelli, G. P., and M. Russo. 1977. Plant virus inclusion bodies. Adv. Virus Res. 21:175–266.

Moyer, J. W., G. G. Kennedy, and L. R. Romanow. 1985. Resistance to watermelon mosaic virus 2 multiplication in *Cucumis melo*. Phytopathology 75:201–205.

Munger, H. M. 1993. Breeding for viral disease resistance in cucurbits. Resistance to Viral Diseases of Vegetables: Genetics and Breeding. Ed. M. M. Kyle. Portland, OR: Timber Press.

Paguio, O. R., H. R. Boerma, and C. W. Kuhn. 1987. Disease resistance, virus concentration, and agronomic performance of soybean infected with cowpea chlorotic mottle virus. Phytopathology 77:703–707.

Paguio, O. R., C. W. Kuhn, and H. R. Boerma. 1988. Resistance-breaking variants of cowpea chlorotic mottle virus in soybean. Plant Dis. 72:768–770.

Ponz, F., and G. Bruening. 1986. Mechanisms of resistance to plant viruses. Ann. Rev. Phytopathol. 24:355–380.

Provvidenti, R. 1993. Sources of resistance to the major virus affecting commonly cultivated cucurbits and strategies for locating and evaluating resistant germplasm. In Resistance to Viral Diseases of Vegetables: Genetics and Breeding. Ed. M. M. Kyle. Portland, OR: Timber Press.

Purcifull, D. E., E. Hiebert, and J. Edwardson. 1984. Watermelon mosaic virus 2. Descriptions of Plant Viruses No. 293. Kew, Surrey, England: Commonw. Mycol. Inst./Assn. Appl. Biol.

Risser, G., M. Pitrat, H. Lecoq, and J.-C. Rode. 1981. Sensibilité variétale du melon (*Cucumis melo* L.) au virus du rabougrissement jaune du melon (MYSV) et à sa transmission par *Aphis gossypii*. Hérédité de la réaction de flétrissement. Agronomie 1:835–838.

Roberts, P. L., and K. R. Wood. 1981. Methods for enhancing the synchrony of cucumber mosaic virus replication in tobacco plants. Phytopath. Z. 102:114–121.

Romanow, L. R., J. W. Moyer, and G. G. Kennedy. 1986. Alteration of efficiencies of acquisition and inoculation of watermelon mosaic virus 2 by plant resistance to the virus and to an aphid vector. Phytopathology 76:1276–1281.

Sarker, S. 1977. Use of protoplasts for plant virus studies. In Methods in

Virology, Vol. 11. Ed. H. Fraenkel-Conrat and R. R. Wagner. New York: Plenum Press.

Seifers, D. L., and T. J. Martin. 1988. Correlation of low level wheat streak mosaic virus resistance in Triumph 64 wheat with low virus titer. Phytopathology 78:703–707.

Smith, C. M. 1989. Plant Resistance to Insects. New York: John Wiley and Sons.

Stoddard S. L., S. A. Lommel, and B. S. Gill. 1987. Evaluation of wheat germplasm for resistance to wheat streak mosaic virus by symptomatology, ELISA, and slot-blot hybridization. Plant Dis. 71:714–719.

Sugimura, Y., and R. E. F. Matthews. 1981. Timing of the synthesis of empty shells and minor nucleoproteins in relation to turnip yellow mosaic virus synthesis in *Brassica* protoplasts. Virology 112:70–80.

Suzuki, N., T. Kudo, Y. Shirako, Y. Ehara, and T. Tachibana. 1989. Distribution of cylindrical inclusion, amorphous inclusion and capsid proteins of watermelon mosaic virus 2 in systemically infected pumpkin leaves. J. Gen. Virol. 70:1085–1091.

Takebe, I. 1977. Protoplasts in the study of plant virus replication. In Methods in Virology, Vol. 11, Ed. H. Fraenkel-Conrat and R. R. Wagner. New York: Plenum Press.

Thompson, D. L., and T. T. Hebert. 1970. Development of maize dwarf mosaic symptoms in eight phytotron environments. Phytopathology 60:1761–1764.

Thresh, J. M. 1984. Progress curves of plant virus disease. Adv. Appl. Biol. 8:1–85.

Travantzis, S. M. 1984. The use of terms for responses of plants to viruses: A reply to recent proposals. Phytopathology 74:379–381.

Van Loon, L. C. 1983. Mechanisms of resistance in virus-infected plants. In The Dynamics of Host Defense. Ed. J. A. Bailey and B. J. Deverall, pp. 123–190. New York: Academic Press.

Zitter, T. A. 1975. Transmission of pepper mottle virus from susceptible and resistant pepper cultivars. Phytopathology 65:110–114.

CHAPTER 10

Interspecific Transfer of Plant Viral Resistance in *Cucurbita*

THEODORE H. SUPERAK, BRIAN T. SCULLY, MOLLY M. KYLE & HENRY M. MUNGER

Introduction: Interspecific Breeding Strategies

Interspecific hybridizations in plant breeding programs are commonly applied to achieve one of two objectives: (1) to develop a unique organism that combines the genomes of both parents; or (2) to transfer a specific gene or trait from a donor species into a recipient species. This is usually followed by a breeding and selection program intended to recover the particular trait in a desired phenotype and restore fertility. Perhaps the best known interspecific hybrid is triticale, an alloploid cross between rye and wheat; but there are numerous horticultural and agronomic crops, derived from interspecific crosses. Interspecific crosses have also been used to develop cytoplasmic male sterility in a number of crop species. Among the squashes, pumpkins, and gourds (*Cucurbita* spp.), interspecific crosses have been employed to serve a diversity of breeding objectives that primarily emphasize the transfer of disease

Dr. Superak is with the Harris Moran Seed Co., Ruskin, FL 33570. Dr. Scully is at the Everglades Research and Education Center, University of Florida, Belle Glade, FL 33430. Dr. Kyle is in the Dept. of Plant Breeding and Biometry, Cornell University, Ithaca, NY 14853. Dr. Munger is in the Departments of Plant Breeding and Vegetable Crops, and pest resistance (Whitaker and Robinson 1986). Additionally, an interspecific F_1 hybrid between two cultivated species, *Cucurbita moschata* and *C. maxima*, is marketed commercially by Japanese seed companies.

Most breeding programs employ a related wild species as the donor parent in backcross or pedigree breeding strategies. The cross may be made directly or may require an intermediate "bridge" species that crosses more readily with the wild species. The success or failure of an interspecific cross is determined by a number of factors. Incompatibility can occur prior to fertilization (presyngamy) or after (postsyngamy). Presyngamic barriers occur during a sequence of pollen-pistil interactions that begin on the pollinated stigma. Initially, the stigmatic surface discriminates among various pollen sources, based on a set of chemical signals between pollen and the stigma (Knox 1986). If germination proceeds, a similar process repeats for the pollen and style. Incompatibility reactions range from a complete failure of pollen germination to distorted or inhibited pollen tube growth in the stylar tissue. These barriers are categorized as incongruity or interspecific incompatibility (Ascher 1986). Incongruity is a passive failure between the pollen and the stigma, whereas interspecific incompatibility is similar to intraspecific incompatibility, and is actively expressed before and after fertilization. Postsyngamic barriers are related to pollen/ovule interactions that include embryonic breakdown, failure of zygote development, abnormal fertilization, and a number of other problems.

In *Cucurbita*, the failure of interspecific crosses is attributed to both pre- and postfertilization barriers (Kwack and Fujieda 1985). A taxonomic review of interspecific relationships within the genus has recently been published by Nee (1990), and cross-compatibility among the cultivated *Cucurbita* taxa and close relatives is summarized in Figure 1. Compatibility is also influenced by the direction of a cross (unilateral incompatibility), cytoplasmic interactions,

---

New York State College of Agriculture and Life Sciences, Cornell University, Ithaca, NY 14853. The authors thank Drs. R. W. Robinson, M. A. Mutschler, and R. Provvidenti for critical reading of the manuscript and Vivianne Scully for assistance with manuscript preparation. Florida Agriculture Experiment Station Journal Series No. R-02148.

and specific genotypic effects. Unilateral incompatibility in *Cucurbita* was identified by Bemis and Nelson (1963), using a complete diallel mating design with ten species. In breeding for viral resistance, deVaulx and Pitrat (1979), and Washek and Munger (1983) found that crosses between the resistant wild species *C. martinezii* and the susceptible species *C. pepo* were most successful when *C. martinezii* (syn. *C. okeechobeensis,* Robinson and Puchalski 1980), was used as the female parent.

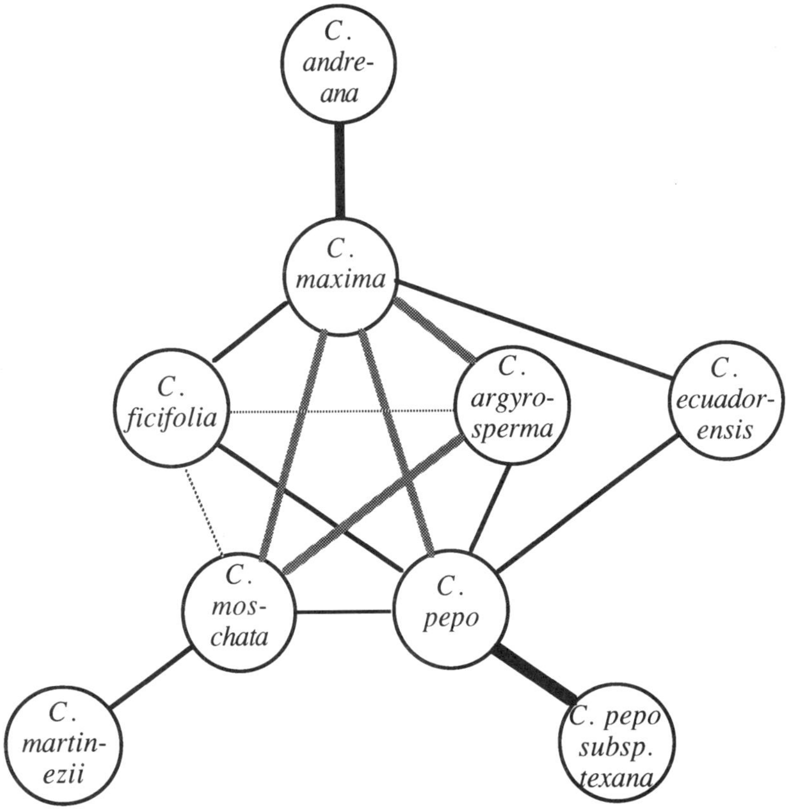

Figure 1. *Cross-compatibility polygon for the cultivated* Cucurbita *and related species.* Thick solid line, *fully fertile hybrids;* medium solid line, *partially fertile hybrids;* thick hatch line, *viable hybrids, self-sterile;* thin solid lines, *sterile hybrids;* thin dotted line, *fruits produced.* C. argyrosperma *is synonymous with* C. mixta *(Merrick and Bates 1989).*

Within a species, unique genetic features of the parents affect the combining ability and success of interspecific crosses. Wall and York (1960) hypothesized that the genetic diversity provided by using an F_1 hybrid as a parent might promote cross-compatibility between *C. pepo* and *C. moschata* 'Butternut'. Their study showed on the contrary that the genotype of the *C. pepo* parent was much more important than the level of heterozygosity. 'Yankee Hybrid', a *C. pepo* F_1 hybrid with little diversity in its parentage, was the most successful female parent, producing 17 viable embryos from 63 pollinations. In contrast, several hundred pollinations with *C. moschata* 'Butternut' used as the male onto *C. pepo* 'Caserta' and four F_1 hybrids with 'Caserta' as one parent gave no viable embryos. The ratio of viable embryos to pollinations was low when the *C. pepo* variety 'Uconn' was the female parent with 'Butternut', improved significantly when ('Uconn' × 'Yankee Hybrid') was the female, and was higher still when 'Yankee Hybrid' was the female. By far the greatest success, 149 embryos from only 27 pollinations, was achieved from the cross ('Uconn' × 'Yankee Hybrid') pollinated by ('Butternut' × 'Golden Cushaw') as male, suggesting that 'Golden Cushaw' itself might be useful as a pollen parent for improving the fertility of the cross.

R. W. Robinson (personal communication) has had the greatest success with 'White Bush Scallop' as the *C. pepo* parent in crosses with *C. moschata*. Crosses between *C. pepo* and the wild gourd *C. ecuadorensis* were more successful when the variety 'Black Jack Zucchini' was used as the *C. pepo* female instead of 'Caserta', 'Scallop', or 'Early Prolific Straightneck' (Robinson and Shail 1987).

A number of methods may be used to overcome interspecific incompatibilities, but only a few have been reported in *Cucurbita* breeding programs. Tissue culture techniques such as embryo culture are most often cited for the production of interspecific F_1 individuals, whereas protoplast fusion, which avoids both pre- and postsyngamic barriers, remains underexploited in *Cucurbita*. Interspecific crosses using a variety of approaches have also been successful in other genera of the Cucurbitaceae, including *Cucumis*, *Lagenaria*, and *Luffa*.

Additional treatments that relax presyngamic barriers include the use of pollen extrusion techniques, repeated pollinations, and mentor pollen, which is a mix of non-viable pollen derived from the female parent and viable pollen from the male (Ascher 1986). Cutting the style to shorten its length and injecting the style with stigmatic exudates can also reduce presyngamic barriers (Ascher 1986). Rejection mechanisms are also controlled with the use of exogenous chemicals, particularly plant growth regulators, and/or the manipulation of temperature and humidity (Kalloo 1987).

After successful pollination, fertilization, and production of an interspecific plant, a number of cytological, genetic, and physiological factors influence sterility of the F_1, F_2, and backcross populations. At this stage, the choice of a breeding method that increases fertility is critical. For example, Pearson et al. (1951) found that *C. maxima* × *C. moschata* amphidiploids provided genetic stability for the hybrid and increased fertility in the F_2 and subsequent generations, thus indicating pedigree breeding as the better approach.

Backcross and Pedigree Methods

Traditionally, interspecific crosses in *Cucurbita* have been followed by backcross and pedigree breeding programs that feature single plants as the selection unit. For simple and oligogenically inherited traits, these methods can rapidly transfer useful levels of resistance into a desired type. Traits that are simply inherited in an intraspecific cross may not be similarly inherited in interspecific crosses. Gene action often changes, and expression of the trait being transferred can be influenced by the genetic background of the recipient species as well as the environment. These changes can include altered inheritance patterns or a possible loss of expression in interspecific populations. Additionally, modifier genes that influence a trait may interact differently when different genomes are combined. Interactions such as epistasis, penetrance, and expressivity are commonly exacerbated in interspecific crosses.

Selfing for one or more generations in a breeding scheme reduces heterozygosity and accelerates the development of inbreds but can also result in the loss of desirable modifier genes. Without a clear understanding of inheritance, population sizes must be adjusted upward to increase the chance of obtaining a desired genotype. With the growing emphasis on the development of hybrids, preference is for dominant, monogenic resistance. Many traits of interest are conditioned by one or a few genes, thus relatively simple inheritance is observed. However, the level of resistance observed in the wild species frequently cannot be reconstituted in breeding lines, presumably due to loss of genes with incremental effects during the recovery of lines with type approaching commercial varieties. Linkage blocks may also play an important role, as free recombination between the genomes does not occur in a number of wide crosses.

Interspecific Bridge Genotypes and Novel Backcross Methods

Backcross methods have been adjusted to reflect interspecific breeding activities directed at the creation and maintenance of interspecific bridges. Fundamental differences exist between the standard and interspecific backcross programs in which a bridge genotype is employed; their joint description as backcross methods is primarily related to similarity of mating schemes rather than genetic properties. Standard backcross procedures practiced within species use a recurrent parent, or set of recurrent parents, and a distinct donor parent. The recurrent parent is first crossed to the donor, then backcrossed sequentially to the offspring. The objective is to transfer a specific gene or trait from the donor and recover a set of individuals near-isogenic with the recurrent parent; however, this is not strictly the case with interspecific backcrossing aimed at the development of genetic bridges.

Whitaker (1956) reported that *C. lundelliana* was cross-compatible with all the cultivated species of *Cucurbita* and suggested its use as a bridge to transfer desirable characters from one species to another. Rhodes (1959) used *C. lundelliana* accessions to combine the germplasm of five *Cucurbita* species (*C. moschata* 'Butternut', 'Long Genoa Queen'; *C. maxima* 'Banana', 'Mammoth Chili', 'Baby Blue'; *C. argyrosperma* 'Green Stripe Cushaw'; *C. lundelliana* accessions; and *C. pepo* 'White Bush Scallop', 'Connecticut Field Pumpkin') into a common gene pool by making two-, three- and four-way crosses, and he gave preliminary information on the possible transfer of bush habit from *C. pepo* and tolerance to powdery mildew (*Sphaerotheca fuliginea*) from *C. lundelliana* and other species by means of the pool (Whitaker 1959). Contin and Munger (1977) reported on the use of *C. moschata* as a bridge species to transfer powdery mildew resistance (PMR) from *C. martinezii* to *C. pepo*. The F_1 of *C. moschata* × *C. martinezii* was used as the male parent in crosses with the interspecific backcross population (*C. pepo* × *C. moschata*) × *C. pepo*. The relatively few plants obtained from this cross could be readily crossed back to *C. pepo* and served to start a successful backcross program for the transfer of PMR (Munger 1990). Washek and Munger (1983) found some segregates with apparent resistance to cucumber mosaic and initiated a backcross program to transfer this to *C. pepo*.

In interspecific backcross programs, multiple species may be used and the role of donor and recurrent parents relaxed. Parent species can alternate recurrently throughout the pedigree, and a donor parent may not be specifically designated (Figure 2). Generally, the purpose of these programs is first to reduce meiotic drive, the nonrandom assortment of chromosomes during meiosis that tends to result in reversion to parental type (Stansfield 1983). Secondly, a genetic bridge to transfer traits between species can be created. The concurrent transfer of an economic trait during the construction of a genetic bridge is a collateral practice that could be integrated into the breeding program. Because no particular parent is continuously recurrent, fully inbred lines and near-isogenic lines are not generated. Three schemes are presented that represent examples of these types of crossing patterns (Figures 2a,b,c).

These bridge backcross schemes have several unique genetic properties. The first cross with the IF_1 defines the majority parent for subsequent odd-numbered generations ($IBC_{1,3}$ . . . etc.), whereas the majority parent for the even-numbered generations is defined by the first alternate recurrent parent (i.e. the grandparent) (Figure 2a). The alternation of parents throughout the crossing procedure influences the additive genetic relationship (a_{ij}) (Falconer 1981;

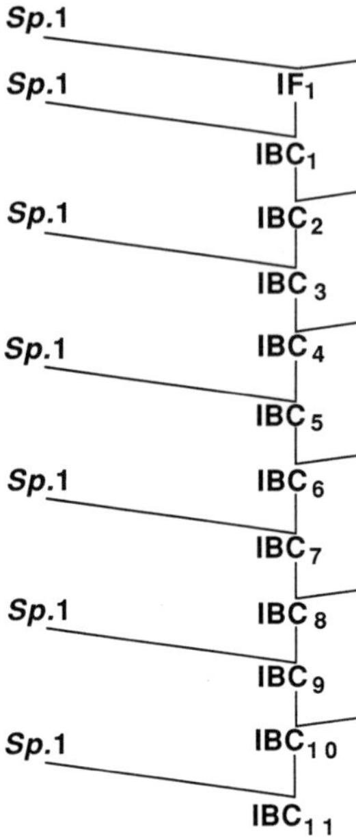

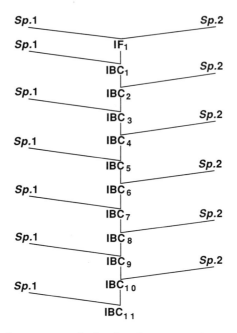

Figure 2a. *The interspecific bridge (congruity) backcross uses two species as alternate recurrent parents.*

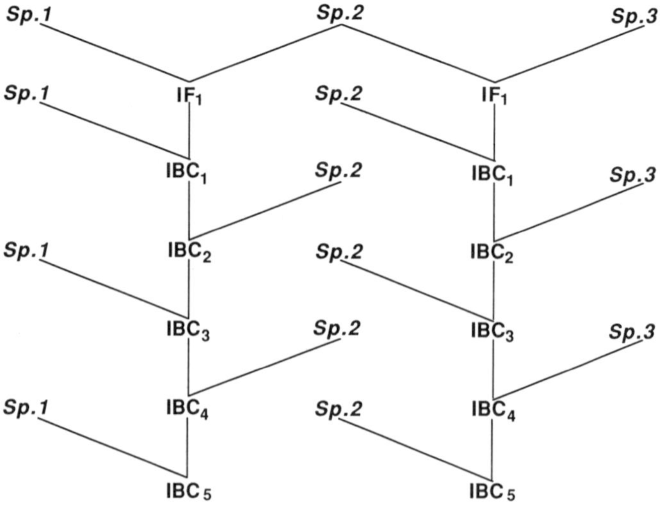

Figure 2b. *An extended bridge backcross scheme that combines overlapping pairs of species as alternate recurrent parents.*

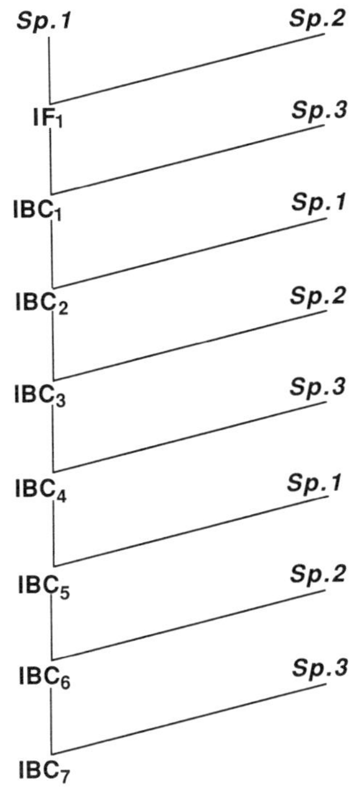

Figure 2c. *A bridge backcross scheme that alternates multiple species as recurrent parents in a single pedigree.*

Wricke and Weber 1986) between parents and progeny and the inbreeding levels (F) of each succeeding backcross generation (Table 1). In a standard backcross procedure, each generation has an increased additive genetic relationship with the recurrent parent and a decreased relationship to the donor parent. When parents are initially non-inbred, the additive relationship values approach 1.0 between recurrent parent and advanced backcross generations; with fully inbred parents this a_{ij} value approaches 2.0. The a_{ij} value approaches 0.0 between donor parent and advanced backcross generations regardless of inbreeding in the parents. In the bridge backcross, odd-numbered generations ($IBC_{1,3}$, etc.) have a_{ij} values that approach ⅔ between the majority parent and advanced backcross generations and ⅓ for the minority parent (Table 1a, Figure 2a). The reverse occurs for the even-numbered generations, assuming that both parents are initially non-inbred. When the parents are fully inbred, the additive relationship values are doubled (Table 1b).

Concurrently, inbreeding levels increase in the bridge backcross procedure, but only marginally. When the parents are non-inbred, IBC generations approach inbreeding values of F=1/6, whereas fully inbred parents produce IBC generations with values that approach F=1/3 (Table 1). This is significantly lower than the standard backcross, which produces backcross generations with inbreeding values of F=1/2 or 1.0 for initially non-inbred or fully inbred parents, respectively.

These computations are predicated on the assumptions usually specified for intraspecific populations, including normal diploid inheritance, independent assortment, known levels of inbreeding, and no gamete/zygote selection. They are intended to provide a guideline rather than a rigorous assessment of genetic relationship. In reality, individual progeny are probably skewed toward the majority parent and reflect selection for traits in addition to increased fertility and cross-compatibility. Nevertheless, the bridge

Table 1. *Additive genetic relationship values* (a_{ij}) *among the parent species* (Sp. 1 *and* Sp. 2) *and the interspecific backcross generations (IBC), and the inbreeding coefficient (F) of each generation of the bridge (congruity) backcross, when: (A) both parents* (Sp. 1 *and* Sp. 2) *are non-inbred (F = 0); and (B) both parents are fully inbred (F = 1).*

2	IF_1	IBC_1	IBC_2	IBC_3	IBC_4	IBC_5	IBC_6	IBC_7	IBC_8	IBC_9	IBC_{10}	$IBC_{\infty\text{-odd}}$	$IBC_{\infty\text{-even}}$
ed Parents													
	$\frac{1}{2}$	$\frac{3}{4}$	$\frac{3}{8}$	$\frac{11}{16}$	$\frac{11}{32}$	$\frac{43}{64}$	$\frac{43}{128}$	$\frac{171}{256}$	$\frac{171}{512}$	$\frac{683}{1024}$	$\frac{683}{2048}$	$\frac{2}{3}$	$\frac{1}{3}$
	$\frac{1}{2}$	$\frac{1}{4}$	$\frac{5}{8}$	$\frac{5}{16}$	$\frac{21}{32}$	$\frac{21}{64}$	$\frac{85}{128}$	$\frac{85}{256}$	$\frac{341}{512}$	$\frac{341}{1024}$	$\frac{1365}{2048}$	$\frac{1}{3}$	$\frac{2}{3}$
	0	$\frac{1}{4}$	$\frac{1}{8}$	$\frac{3}{16}$	$\frac{5}{32}$	$\frac{11}{64}$	$\frac{21}{128}$	$\frac{43}{256}$	$\frac{85}{512}$	$\frac{171}{1024}$	$\frac{341}{2048}$	$\frac{1}{6}$	$\frac{1}{6}$
arents													
	1	$\frac{3}{2}$	$\frac{3}{4}$	$\frac{11}{8}$	$\frac{11}{16}$	$\frac{43}{32}$	$\frac{43}{64}$	$\frac{171}{128}$	$\frac{171}{256}$	$\frac{683}{512}$	$\frac{683}{1024}$	$\frac{4}{3}$	$\frac{2}{3}$
	1	$\frac{1}{2}$	$\frac{5}{4}$	$\frac{5}{8}$	$\frac{21}{16}$	$\frac{21}{32}$	$\frac{85}{64}$	$\frac{85}{128}$	$\frac{341}{256}$	$\frac{341}{512}$	$\frac{1365}{1024}$	$\frac{2}{3}$	$\frac{4}{3}$
	0	$\frac{1}{2}$	$\frac{1}{4}$	$\frac{3}{8}$	$\frac{5}{16}$	$\frac{11}{32}$	$\frac{21}{64}$	$\frac{43}{128}$	$\frac{85}{256}$	$\frac{171}{512}$	$\frac{341}{1024}$	$\frac{1}{3}$	$\frac{1}{3}$

backcross produces two types of interspecific genotypes, each with a different majority parent. These genotypes can be further intermated to provide the breeder with a diverse germplasm base to facilitate interspecific transfer.

Bridge backcross schemes can also be linked to include the use of multiple parent species (Figure 2b). This extended version of the bridge backcross features parallel mating schemes that share a common parent (Species 2) but also isolates two parental species (Species 1 and 3) that may be incompatible. The species common in both crossing schemes should have broader compatibility relative to the other taxa. In *Cucurbita*, *C. lundelliana* and *C. moschata* have commonly served as intermediate species for the transfer of economic traits. Such schemes might aid in the transfer of traits from genetically isolated species, such as the xerophytic *Cucurbita* group.

Another possible variation of the bridge backcross includes the use of three alternate recurrent parent species (Figure 2c) instead of just two. Three-way crosses in *Cucurbita* have been successful with *C. lundelliana* and any two cultivated species (Rhodes 1959), and with *C. martinezii*, *C. pepo*, and *C. moschata* (Whitaker and Robinson 1986). These types of crosses were usually followed by pedigree breeding methods, but the use of a backcross scheme may allow the development of bridge genotypes with more universal cross-compatibility. Clearly, these extended bridge backcross schemes, e.g. Figure 2c, can be modified to meet the particular needs of a breeding program that includes related species as a genetic resource for disease resistance.

Variations of these novel backcross schemes have proven useful in several vegetable breeding programs. By recurrently alternating parents in a crossing scheme between common beans (*Phaseolus vulgaris*) and tepary beans (*P. acutifolius*), Haghighi and Ascher (1988) developed a genetic bridge with increased pollen viability and seed fecundity. They called this basic procedure a "congruity backcross" and denoted the backcross generations as "CBC." With some minor variation, this procedure was independently developed in *Cucurbita* at the Harris Seed Co., Rochester, NY with *C. pepo* and *C. moschata* as the parent species (Figure 4). Because the stated purpose of both the *Phaseolus* and *Cucurbita* programs was to generate interspecific bridge genotypes, we feel the term "interspecific bridge backcross" more clearly describes the objectives. The program commences with an interspecific cross between two related taxa followed by a first backcross to one of the parent species (Figures 2, 3, and 4). Subsequent backcrosses can alternate parent species recurrently until the desired level of fertility is attained. This program is essentially an alternate grandparent backcross (Figure 2a).

Another modification of standard F_1 bridging and backcross procedures uses the interspecific hybrid (IF_1) as the recurrent

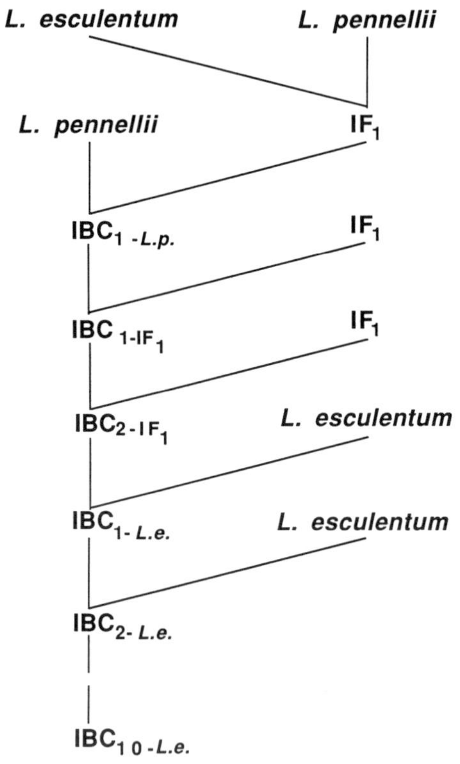

Figure 3. *A scheme used by Mutschler to develop an interspecific bridge for the transfer of cytoplasm from* L. pennellii *to* L. esculentum. *The interspecific F_1 (IF_1) served as the recurrent male parent for two gernerations and was followed by* L. esculentum *in the remaining crosses. The subscript designation following the IBC notation indicates the recurrent parent of that backcross.*

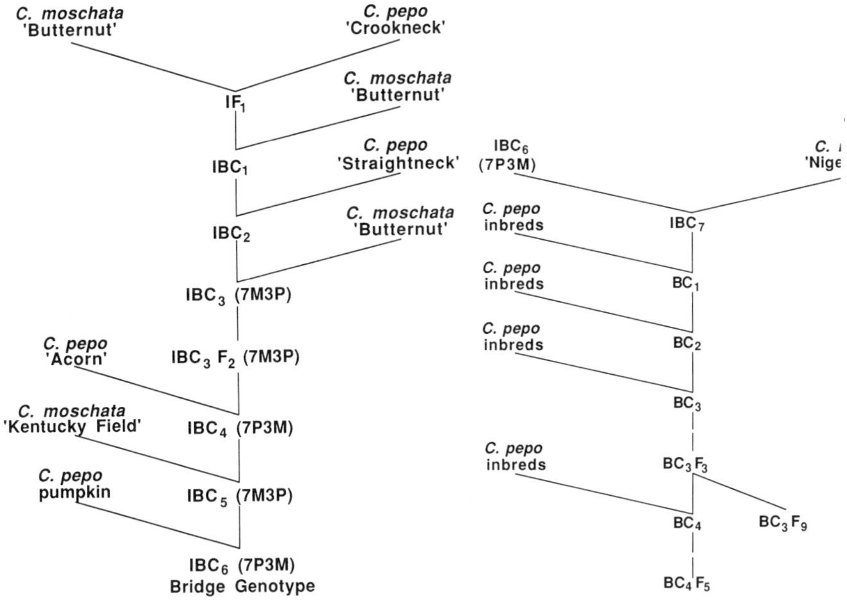

Figure 4. *Pedigree for the development of a genetic bridge between* Cucurbita pepo *and* C. moschata *followed by the backcross program used to transfer resistance to zucchini yellow mosaic virus from* Cucurbita moschata *'Nigerian Local' across the bridge genotype (7P3M) into* C. pepo.

parent (Figure 3). This scheme was developed by M. A. Mutschler (personal communication) to generate an interspecific bridge for the transfer of cytoplasm from *L. pennellii* Corr. to *L. esculentum* Mill. In this case, the IF_1 was used as the male parent in the initial interspecific backcross (IBC_1) with *L. pennellii*. Fertility was recovered in the $IBC_{2\text{-}L.e.}$ generation. Progeny derived from the cross (*L. pennellii* × IF_1) have a close relationship to the wild species, but the additional crosses with the IF_1 reduce the coancestry between the backcross progeny and *L. pennellii*. In the next two backcrosses, use of *L. esculentum* as the male was impossible. Using *L. esculentum* as the female, 50 to 100 pollinations were required to obtain any seed or culturable embryos. Embryo culture was required in the backcrosses involving the IF_1 and in the first backcross to *L. esculentum*.

In a scheme using recurrent crosses to the IF_1, the level of inbreeding will continue to approach ½. This procedure produces plants that are fertile, partially inbred, and that can function as an interspecific bridge genotype without a specified majority parent. The scheme was used to circumvent unilateral incompatibility by increasing the proportion of the *L. esculentum* genome in the successive backcross generation until *L. esculentum* pollen was not rejected by the stigma (Figure 3). Backcrosses to the IF_1 result in a slower approach to the *L. esculentum* type than straight backcrosses to *L. esculentum*, but when the latter cross is not feasible, this is a viable alternative strategy.

Decisions relative to the direction of a cross and variations in compatibility of genotypes among the parental species cannot be generalized. Previous experience and imagination are important elements in these judgments. After the first interspecific cross, four mating combinations are possible, including the use of IF_1 as male or female with either parental species. In *Cucurbita*, experience suggests that where possible, a common cytoplasm throughout the crossing procedure is generally best. Taxa such as *C. moschata* and *C. maxima* have cross-compatible cytoplasms, and *C. moschata* is compatible with that of *C. pepo*.

A Case Study: Interspecific Transfer of ZYMV Resistance from *C. moschata* to *C. pepo*

Zucchini yellow mosaic virus (ZYMV) is an aphid-transmitted potyvirus. The first reports of its occurrence date from the early 1980s, when ZYMV was reported from countries near the Mediterranean Sea (Lecoq et al. 1981; Lisa et al. 1981; Provvidenti et al. 1984). In 1982 and 1983, outbreaks of ZYMV seriously affected *Cucurbita* production in California, Florida, and the Northeastern USA (Provvidenti et al. 1984). Infected plants exhibited foliar distortion, bright yellow mosaic, and necrosis, and if fruit was produced it was small and highly distorted with knobs and color break (Provvidenti et al. 1984). Provvidenti et al. (1984) classified two distinct pathotypes, ZYMV-CT (isolate CT 82-35) from Connecticut, which is the more severe strain, and ZYMV-FL (isolate FL-H-82-50) from Florida. These two pathotypes have slightly different geographic distributions and incite different symptoms. ZYMV-CT caused significantly brighter yellow foliage, similar to the symptoms described from Italy, and was also isolated from infected field plantings in California. ZYMV-FL was additionally found on cucurbits in New York and California.

Resistance to ZYMV was reported in *C. ecuadorensis* and *C. moschata* 'Nigerian Local' (Provvidenti et al. 1984). Robinson et al. (1988) crossed the susceptible *C. maxima* 'Buttercup' to *C. ecuadorensis* and determined that resistance to ZYMV-CT was conferred by a single co-dominant gene designated *zym*. The inheritance of resistance to both strains of ZYMV in *C. moschata* was identified as a single dominant gene (Munger and Provvidenti 1987, Provvidenti 1988). Apparently this same gene confers resistance with partially dominant gene action in crosses between *C. pepo* and *C. moschata* (Munger and Provvidenti 1987). Efforts to transfer resistance from 'Nigerian Local' to butternut-type squashes is currently under way through use of standard backcross procedures (Munger and Provvidenti 1987).

The purpose of the breeding program presented as an illustration of interspecific transfer of viral resistance via a bridge genotype was to (1) develop a genetic bridge between *C. moschata* and *C. pepo*, and (2) use the bridge genotypes for the orderly and rapid deployment of desired traits from *C. moschata* into the elite *C. pepo* inbred lines needed for hybrid variety production.

Efforts to construct the genetic bridge between *C. moschata* and *C. pepo* were initiated in 1975 at the Harris-Moran Seed Co. with the bridging backcross technique. Isolation blocks of *C. moschata* 'Butternut' and *C. pepo* 'Summer Crookneck', which cross-pollinate infrequently, were grown in adjacent rows in New York. Subsequent plantings of the *C. moschata* breeding line revealed that a spontaneous outcross between the two species had produced several viable plants. Male flowers were removed from these interspecific hybrids (IF_1) and the female flowers pollinated with *C. moschata* to produce the first interspecific backcross generation (IBC_1) (Figure 4). Pollen parents were then alternated between *C. pepo* and *C.*

moschata until IBC_3. The IBC_3 generation was selfed, and following generations were backcrossed as males to the alternating recurrent parents. Additionally, the *C. pepo* parental varieties were changed to include an acorn squash and pumpkin. The IBC_6 generation was available in 1984, with a theoretical genetic composition of 0.336 ($43/128 \simeq 1/3$) *C. pepo* and 0.664 ($85/125 \simeq 2/3$) *C. moschata* genome (Table 1a). The change in parental genotype was an endeavor to improve fertility in succeeding generations. Despite this effort, after several generations of alternating backcrosses, fertility and cross-compatibility found in succeeding generations had improved but was still well below that found within intraspecific crosses. The improved levels of fertility and cross-compatibility reported in succeeding generations of *P. vulgaris* by *P. acutifolius* crosses by Haghighi and Ascher (1988) were far superior.

When resistance to ZYMV was identified by R. Provvidenti in 'Nigerian Local', seed was obtained in 1984 and a breeding program was initiated to transfer resistance into *C. pepo*. 'Nigerian Local' was first crossed as a male to the bridge genotype designated IBC_6 (this genotype has been designated 7P3M indicating the sixth interspecific backcross generation $\simeq 2/3$ *C. pepo*: $\simeq 1/3$ *C. moschata*) in Costa Rica during the spring of 1985. Viable F_1 seed and fertile F_1 plants were produced. This generation proportionally had ⅔ *C. moschata* and ⅓ *C. pepo* genomes. This cross was followed by a backcross program that transferred dominant gene resistance into a set of recurrent parent lines, including zucchini, striped, grey, white, and yellow summer squash. These inbred lines were selected for horticultural type, superior combining ability, disease resistance, and representative genetic diversity across the summer squash group.

The first backcross generation (BC_1) was produced in the summer of 1985 in New York, and theoretically was composed of ⅔ *C. pepo* and ⅓ *C. moschata* genomes. To improve the certainty of success, crosses were made bidirectionally, and embryo rescue techniques were employed in the BC_1 and BC_2 generations. Hindsight has suggested that these precautions may have been unnecessary. Starting with BC_1 in the fall of 1985, all subsequent backcross and selfed generations were mechanically inoculated with ZYMV-CT (isolate CT 82-35), using the petiole technique of Provvidenti et al. (1978a,b). The second and third backcross generations were also produced with bidirectional crosses in the fall and spring of 1986 and 1987, respectively, and recovered ≃5/6 of the *C. pepo* genome. The BC_1 and BC_2 generations had a high level of resistance and a large proportion of resistant individuals. The BC_3 generation had a low proportion of resistant individuals which was attributed to the use of a more virulent strain of ZYMV derived from infected plants from previous generations.

Resistant individuals in the BC_3 F_3 generation were crossed again to summer squash inbreds, advanced to BC_4 F_5, and ZYMV resistant lines were identified (Figure 4). Additional pedigree selection was applied from BC_3 F_3 to the BC_3 F_9 generation when homozygous resistant lines were identified and subsequently evaluated for hybrid production. These first trial hybrids with grey, white, and zucchini types were evaluated in the fall of 1988 in Florida under ZYMV-FL pressure. Resistant inbreds, susceptible standards, and hybrids were included in this test. Hybrids with one resistant parent showed mild plant and fruit symptoms, whereas the susceptible checks were severely stunted, and resistant lines were nearly symptomless. This suggested the need for resistance in both parents for production of hybrids with satisfactory resistance. A problem encountered along the way has been the apparent semi-independence of resistance to plant and fruit symptoms. Fruit on a plant apparently resistant to ZYMV-CT will have a bumpy surface, despite significantly reduced foliar symptoms. The Florida strain behaves similarly, but fruit symptoms are generally milder. Another, less troublesome, problem has been sterility or lack of male flowers in some lines.

Conclusions

Interspecific backcrossing schemes have been developed and applied in several genera including examples in three vegetable crops, *Phaseolus*, *Lycopersicon*, and *Cucurbita*. The genus *Cucurbita* is a large and widely adapted genus with only a few important cultivated species. The cultivated species have been improved by the interspecific transfer of traits, especially disease and pest resistance. The degree of difficulty associated with this task depends on how closely the species are related and the development of techniques to augment interspecific hybridization. A number of approaches have been used to overcome interspecific barriers that range from sampling a number of genotypes within a species to protoplast fusion.

Breeding programs that routinely incorporate interspecific crosses have commonly modified standard backcross methods to better achieve their breeding objectives. The novel backcross methods are one set of schemes that are similar to standard backcross methods, but in practice they actually have few properties in common. The purpose of these procedures is to develop fertile interspecific bridge genotypes that can be used repeatedly for the transfer of economically important traits. The bridge backcross method was one of the novel backcross schemes used successfully to create a bridge between *C. moschata* to *C. pepo.* It was subsequently employed for the transfer of resistance to zucchini yellow mosaic virus from *C. moschata* to *C. pepo*. The approaches that have been developed and successfully applied in *Cucurbita* to transfer viral resistance from wild species to cultivated taxa where resistance was not available are generally applicable when interspecific breeding strategies are necessary.

Literature Cited

Ascher, P. D. 1986. Incompatibility and incongruity: Two mechanisms preventing gene transfer between taxa. In: Biotechnology and Ecology of Pollen: Proceedings International Conference on Biotechnology and Ecology of Pollen. Ed. D. L. Mulcahy, G. B. Mulcahy, and E. Ottoviano. pp. 251–256. Berlin: Springer-Verlag.

Bemis, W. P., and J. M. Nelson. 1963. Interspecific hybridization within the genus *Cucurbita*: I. Fruit set, seed and embryo development. J. Ariz. Acad. Sci. 2:104–107.

Contin, M., and H. M. Munger. 1977. Inheritance of powdery mildew resistance in an interspecific cross with *Cucurbita martinezii*. HortScience 12:397.

de Vaulx, R. D., and M. Pitrat. 1979. Interspecific cross between *Cucurbita pepo* and *C. martinezii*. Cucurbit Gen. Coop. 2:35.

Falconer, D. C. 1981. Introduction to Quantitative Genetics (Second Ed.). New York: Longmans Grp. Ltd.

Haghighi, K. R., and P. D. Ascher. 1988. Fertile, intermediate hybrids between *Phaseolus vulgaris* and *P. acutifolius* from congruity backcrossing. Sex. Plant Reprod. 1:51–58.

Kalloo, D. 1987. Vegetable Breeding. Boca Raton, FL: CRC Press, Inc.

Knox, R. B. 1986. Pollen-pistil interactions. In Cellular Interactions: Encyclopedia of Plant Physiology, New Series #17. Ed. H. F. Linskens and J. Hyslop-Harrison. pp. 508–608. Berlin: Springer-Verlag.

Kwack, S. N., and K. Fujieda. 1985. Breeding high female lines through interspecific hybridization of *Cucurbita*. Cucurbit Genet. Coop. 8:78.

Lecoq, H., M. Pitrat and M. Clement. 1981. Identification et caracterization d'un potyvirus provoquant la maladie du rabougrissement jaune du melon. Agronomie 1:827–834.

Lisa, V., G. Boccardo, G. D'Agostino, G. Dellavalle, and M. D'Aquino. 1981. Characterization of a potyvirus that causes zucchini yellow mosaic. Phytopathology 71:668–672.

Merrick, L. C., and D. M. Bates. 1989. Classification and nomenclature of *Cucurbita argyrosperma*. Baileya 23:94–102.

Munger, H. M. 1990. Availability and use of interspecific populations involving *Cucurbita moschata* and *C. pepo*. Cucurbit Genet. Coop. 13:49.

Munger, H. M., and R. Provvidenti. 1987. Inheritance of resistance to ZYMV in *Cucurbita moschata*. Cucurbit Genet. Coop. 10:80–81.

Nee, M. 1990. The domestication of *Cucurbita* (Cucurbitaceae). Econ. Bot. 44:56–68.

Pearson, O. H., R. Hopp, and G. W. Bohn. 1951. Notes on species crosses in *Cucurbita*. Proc. Amer. Soc. Hort. Sci. 57:310–322.

Provvidenti, R. 1988. Viral diseases and genetic sources of resistance in *Cucurbita* species. In Biology and Utilization of the Cucurbitaceae. Ed. D. M. Bates, R. W. Robinson, and C. Jeffrey. pp. 427–435. Ithaca, NY: Cornell Univ. Press.

Provvidenti, R., D. Gonsalves, and H. S. Humaydan. 1984. Occurrence of zucchini yellow mosaic virus in cucurbits from Connecticut, New York, Florida and California. Plant Dis. 68:443–446.

Provvidenti, R., R. W. Robinson, and H. M. Munger. 1978a. Multiple virus resistance in *Cucurbita* species. Cucurbit Genet. Coop. 1:26–27.

Provvidenti, R., R. W. Robinson, and H. M. Munger. 1978b. Resistance in feral species to six viruses infecting *Cucurbita*. Plant Dis. Rept. 62:326–329.

Rhodes, A. M. 1959. Species hybridization and interspecific gene transfer in the genus *Cucurbita*. Proc. Amer. Soc. Hort. Sci. 74:546–551.

Robinson, R. W., and J. T. Puchalski. 1980. Synonymy of *Cucurbita martinezii* and *C. okeechobeensis*. Cucurbit Genet. Coop. 3:45–46.

Robinson, R. W., and J. W. Shail. 1987. Genetic variability for compatibility of an interspecific cross. Cucurbit Genet. Coop. 10:88.

Robinson, R. W., N. F. Weeden, and R. Provvidenti. 1988. Inheritance of resistance to ZYMV in interspecific cross, *Cucurbita maxima* × *C. ecuadorensis*. Cucurbit Genet. Coop. 11:74–75.

Sowell, G., and W. L. Corlley. 1973. Resistance of *Cucurbita* plant introductions to powdery mildew. Hortscience 8:492–493.

Stansfield, W. D. 1983. Genetics. Schaum's Outline Series. 2nd edition. New York: McGraw Hill.

Wall, J. R., and T. L. York. 1960. Gametic diversity as an aid to interspecific hybridization in *Phaseolus* and *Cucurbita*. Proc. Amer. Soc. Hort. Sci. 75:419–428.

Washek, R. L., and H. M. Munger. 1983. Hybridization of *Cucurbita pepo* with disease resistant *Cucurbita* species. Cucurbit Genet. Coop. 6:92.

Whitaker, T. W. 1956. The origin of cultivated *Cucurbita*. Am. Naturalist 90:171–176.

Whitaker, T. W. 1959. An interspecific cross in *Cucurbita* (*C. lundelliana* Bailey × *C. moschata* Duch.). Madrono 15:4–13.

Whitaker, T. W., and R. W. Robinson. 1986. Squash Breeding. In Breeding Vegetable Crops. Ed. M. J. Basset. Westport, CT: AVI Publ. Co. pp. 204–242.

Wricke, G., and W. E. Weber. 1986. Quantitative Genetics and Selection in Plant Breeding. Berlin, Germany: Walter de Gruyter and Co.

CHAPTER 11

Mechanisms of Plant Viral Pathogenesis

CANDACE W. COLLMER & STEPHEN H. HOWELL

Introduction and Terminology

In the simplest terms, viral pathogenesis results when plant resistance fails. More accurately, plant viral *pathogenesis* represents one possible outcome of the complex interplay among processes within a plant involved in viral replication, pathogenicity, and host plant resistance. By the definitions of R. S. S. Fraser (1985), a plant that supports the replication of a particular virus is a *host* for that virus and thereby is *susceptible* to it. Such an interaction between plant host and viral parasite may or may not be a pathogenic one resulting in *disease*, expressed as host damage measured by symptom severity or yield loss. Pathogenesis is a visible result of multiple interactions between viral and host components as the virus uncoats, replicates, assembles, and spreads throughout the plant. Replication of a virus within a plant does not always lead to pathogenesis, however, and the amount of disease "may be at least partly distinct in mechanistic terms from the amount of multiplication of the pathogen" (Fraser 1985). Plant breeders in particular are familiar with the concept of *tolerance*, whereby a host plant supports viral replication without concomitant visible or severe disease. This potential uncoupling of viral replication and viral pathogenesis has led to a rather broad

Dr. Collmer is in the Department of Biology, Wells College, Aurora, NY 13026. Dr. Howell is at the Boyce Thompson Institute for Plant Research, Cornell University, Tower Road, Ithaca, NY 14853.

definition of *resistance*, i.e. as inhibition of any stage of the viral life cycle *or* of the development of pathogenic effects in the host. Thus, resistance may operate mechanistically against viral replication or spread, against the development of disease, or against both. One specific type of resistance occurs at the initial contact between virus and plant, resulting in necrotic spots indicative of a *hypersensitive response* that prevents viral spread and, therefore, pathogenesis.

How viruses induce disease symptoms is a question about which little is known, and several simple observations suggest that the answer(s) may be anything but simple. First, as mentioned above, the amount of virus present in a plant does not always correlate with the severity of disease symptoms. In fact, two viral, subviral, or viroid pathogens that are quite closely related and that replicate with similar efficiency in their plant host may induce radically different disease patterns. Second, two viruses from dissimilar taxonomic groups may produce quite similar symptoms, as indicated by the large number of plant viruses bearing the name "mosaic." Finally, and of most familiarity to plant breeders, the same virus on two plants differing only by the presence or absence of a particular resistance gene can produce completely different symptoms.

Whereas the second of these observations suggests commonalities among "unrelated" viruses in the induction of certain symptoms, the first suggests the existence of some unique virus-host interactions and disease pathways. The third asserts the undeniable involvement of the host plant in pathogenesis. Taken together, the apparent complexities of viral pathogenesis underscore the potential power of some of the newly developed molecular techniques that allow the precise manipulation of a single component of such a system, usually a single gene of plant, viral, or subviral origin. As we discuss below, some experimental studies of this type are starting to yield interesting results.

Approaches to the Puzzle of Viral Pathogenesis

Although the puzzle of how viruses induce disease symptoms in plants has attracted the attention of plant virologists for over a century, only recently has progress been made toward identifying molecular mechanisms of disease development. Matthews (1973) divided such efforts, historically, into three phases—"The Symptom Era," from about 1900–1935, during which hundreds of diseases were described and named; "The Era of Prejudice," from about 1936–1965, when the description of virus symptoms fell into disrepute in favor of molecular studies of the viruses themselves; and the era since about 1965, a time of synthesis. The latter era drew together the knowledge accumulated to date and progressed forward with the application of sensitive analytical techniques and the wider use of electron microscopy. The development of recombinant DNA and transgenic plant technologies has opened a new chapter in understanding viral pathogenesis. The need to exploit new opportunities is apparent when one considers that even the simplest viral system is likely to involve several viral-specified RNAs and a minimum of 4–6 viral-encoded proteins, all of which may interact with host-encoded components during the viral life cycle. Even in the simplest scenario, in which one viral gene product is primarily responsible for symptom production, any other viral component that affects the level of viral replication, movement, etc. would also indirectly affect the level of that gene product and thus the extent of its perturbation of the plant. In addition, it is quite likely that there are several determinants of any symptomatology, involving both viral nucleic acids and proteins. This may explain why earlier attempts to map determinants of symptom production to specific plant viral RNAs or genes, mostly through the construction of pseudorecombinants of viruses with multicomponent genomes (artifically constructed mixtures in which the individual genomic RNAs are derived from two different, but related, viruses; see Fraser 1987), have been generally unsatisfying.

It follows that experimental systems likely to prove most fruitful in the study of viral pathogenesis are those in which pathogenesis and viral replication are separable. Examples of systems of particular promise are discussed below.

Plant Genes that Suppress Viral Pathogenicity

Given the involvement of the host plant in viral pathogenesis, the study of plant genes implicated in the suppression of viral pathogenicity is certain to be fruitful. Much work has been and continues to be done on this topic in several fields, including plant breeding, and its review is beyond the scope of the present chapter. It seems important to mention, however, that certain experimental systems appear particularly promising for the study of the process of viral pathogenicity, specifically, those in which the effect of a resistance gene on viral pathogenesis is separable from its effect on viral replication. One example of such a system is the *Tm-1* gene in tomato, which confers resistance to tobacco mosaic virus (TMV). *Tm-1* has pleiotropic effects on symptom suppression and inhibition of viral replication. The gene is dominant for the suppression of mosaic symptoms caused by the virus, but inhibition of TMV replication is gene-dosage dependent. Most interesting, the two effects of the *Tm-1* gene are separable at elevated temperature. Two sets of tomato plants (susceptible versus resistant) that are near-isogenic for all but *Tm-1* support the same level of TMV replication when grown at 33°C, yet symptoms that are apparent in the +/+ (susceptible) plants are suppressed in the *Tm-1* (resistant) plants (Fraser and Loughlin 1980). Such observations beg for the isolation of the *Tm-1* gene product in order to study its role in inhibiting viral pathogenesis, a feat under way (see Young et al. 1993, this volume) but not yet accomplished.

Defining Determinants of Pathogenicity in Viral and Subviral Pathogens

Some progress has been made in teasing apart determinants of viral pathogenicity versus viral replication in the viral pathogen, particularly with two systems of special relevance to work in our own laboratory. First, work on small subviral pathogenic RNAs, specifically viroids and satellite RNAs, has succeeded in pinpointing specific domains in which sequence changes can affect pathogenicity but not replication. Second, techniques of "reverse genetics" with the RNA virus TMV, and the construction of transgenic plants containing an isolated gene of the DNA virus cauliflower mosaic virus (CaMV), have shown a definite connection between structural proteins of these viruses and induction of pathogenesis. In the remainder of this chapter we present progress from selected studies in these two areas of plant viral pathogenesis. Although this overview is not meant to be inclusive, it should be representative of current efforts to define plant viral and subviral determinants of pathogenicity at a molecular level. A review from an alternative viewpoint (Fraser 1987) has described the wealth of data collected previously on the pathogenic effects of viruses on plant growth, development, and symptom production.

Mapping Pathogenesis Domains in Small Pathogenic RNAs

The natural occurrence within two classes of small pathogenic RNAs of closely related isolates causing disease symptoms of varying severity has provided fertile ground for molecular studies seeking to correlate RNA structure with pathogenicity. For both potato spindle tuber viroid (PSTV) and the parasitic satellite RNAs shown to modify the disease symptoms of cucumber mosaic virus (CMV) and turnip crinkle virus (TCV), progress has been made in defining specific domains of the RNA molecules involved in symptom production.

Potato Spindle Tuber Viroid

A viroid, a small, single-stranded, circular RNA molecule that lacks a capsid protein, can be a destructive plant pathogen. In PSTV, comparison of the nucleotide sequences of eight field isolates ranging in virulence on tomato from mild to lethal revealed that the differences in pathogenicity could be directly correlated with the structural stability of a particular region of the molecule denoted the virulence-modulating (VM) region (Schnölzer et al. 1985). Because the calculated melting temperature of the VM region decreases with increasing virulence of the PSTV isolate, it was hypothesized that differences in disease severity might be related to accessibility of that particular region of PSTV for binding to an unknown host factor. Later work from the same laboratory offered a possible primary cellular target with which that region of the viroid molecule might interact and thus incite disease. The "lower strand" of the VM region is complementary to the tomato 7S RNA (Haas et al. 1988), a finding that suggests that the viroid might form a base-paired region with this putative component of a plant signal recognition particle and somehow disrupt its normal function in protein translocation. More recent work (Harders et al. 1989), however, precisely localizing the majority of PSTV (+) strand and (−) strand sequences within the nucleoli of infected cells, argues against their pathogenic role in the cytoplasm. Although various alternative scenarios for disruption of normal cellular functions can be imagined, an actual interaction between PSTV and any host plant RNA remains to be demonstrated.

Satellite RNAs Associated with Cucumber Mosaic Virus

Satellite RNAs are RNA molecules that are dependent on a helper virus for replication and contain appreciable nucleotide sequence distinct from that of the helper virus. The satellite RNAs associated with a variety of plant viruses have recently attracted considerable attention because of the existence of many naturally occurring sequence variants with differing phenotypes and their ability to modulate virally induced plant disease (for a review see Kaper and Collmer 1988). Of particular interest has been the ability of some, but not other, closely related sequence variants of the satellite RNAs of cucumber mosaic virus to produce lethal necrosis in tomato when co-inoculated with the helper virus. Whether or not a particular satellite is necrogenic cannot be explained by differences in satellite replication in the plant host. Attempts to correlate changes in nucleotide sequence with pathogenicity of 20 different variants did not identify domains of the satellite RNA responsible for induction of necrosis (Kaper et al. 1988). Independent work in four laboratories involving the construction of chimeras between necrosis-inducing and attenuating satellites, however, localized the necrosis domain in the 3′ half of the molecule (Devic et al. 1989; Kurath and Palukaitis 1989; Masuta and Takanami 1989; C. W. Collmer and J. M. Kaper unpublished, cited in Daubert 1988). Although the comparison (Jacquemond and Lauquin 1988) of the nucleotide sequences of two very closely related French satellites that do and do not induce necrosis had suggested that a very few changes in nucleotide sequence could dramatically alter the satellite's phenotype, Masuta and Takanami (1989) actually demonstrated that the alteration of a single nucleotide, from a C to a T in an expressible cDNA clone of a necrogenic satellite, was sufficient to change the phenotype from necrogenic to attenuating. Viewed as a whole, these data suggest that a particular secondary or tertiary structure at the very 3′ end of the satellite RNA may be the determinant of necrosis induction in tomato. Parallel studies have localized the domain involved in a different phenotype, chlorosis induction in tomato or tobacco (depending on the particular satellite RNA), to the 5′ half of those satellites (Devic et al. 1989; Kurath and Palukaitis 1989; Masuta and Takanami 1989).

Although it has been suggested (Masuta and Takanami 1989) that alternative secondary or tertiary RNA structures and their resultant differential interactions with host components might control disease phenotypes, what such host components might be is presently unknown. In the case of the necrogenic phenotype, the observation that at least some of the satellite RNAs that are lethal in tomato have an alternative disease-attenuating effect in tobacco and most other hosts (Kaper and Waterworth 1977) may be exploited in the search for such putative host components. Data from several laboratories has shown, however, that the presence of a particular satellite RNA in a particular host plant is insufficient to induce a particular phenotype. Changing the strain of the helper virus (CMV) supporting the satellite RNA can completely change the symptoms of the disease (Masuta et al. 1988; Palukaitis 1988; Kaper et al. 1990). Thus, in this system, viral pathogenesis appears to result from a trilateral interaction among helper virus, satellite RNA, and host plant.

Small RNAs Associated with Turnip Crinkle Virus

A second system of particular promise for exploring mechanisms of pathogenesis is the turnip crinkle virus (TCV) family of small, replication-dependent RNAs (reviewed in Simon 1988). The John Innes isolate of TCV (and the derivative Massachusetts isolate, TCV-M) supports the replication of a group of related satellite RNAs. The avirulent satellites D (194 bases) and F (230 bases) are closely related and show no appreciable nucleotide sequence homology with helper viral RNA; they are dependent on TCV for replication but have no apparent effect on TCV-induced symptoms (Altenbach and Howell 1981; Simon and Howell 1986). The virulent satellite C (355 bases), on the other hand, intensifies the leaf crinkling associated with TCV in several cruciferous hosts (Li and Simon 1990). Satellite RNA C is an unusual hybrid molecule composed of a 3′ domain (166 bases) very similar to two regions of the 3′ end of TCV

(Carrington et al. 1987), and a 5′ domain (189 bases) similar to the entire sequence of the avirulent satellite (sat)-RNA D (Simon and Howell 1986). A chimeric molecule constructed experimentally from 155 bases at the 5′ end of sat-RNA F joined to 200 bases from the 3′ end of sat-RNA C also intensified TCV symptoms, thereby mapping the region determining virulence to the 3′ 200 bases of sat-RNA C (Simon et al. 1988). Because this region responsible for symptom intensification is homologous to the 3′-untranslated region of TCV RNA, it has been suggested that the 3′ end of the TCV genome itself produces mild symptoms in plants infected with TCV alone, and that the intensification of symptoms by sat-RNA C is due simply to the increased dosage of this 3′ region in those plants (Simon and Howell 1986). That the additional presence of sat-RNA C has no effect on host phenotype in several cruciferous hosts in which infection by TCV alone produces no symptoms is consistent with this hypothesis (Li and Simon 1990).

In addition, a second type of small RNA molecule sometimes associated with, and dependent upon, the Berkeley strain of TCV (TCV-B) has been shown (Li et al. 1989) to contain the same 3′ region as TCV RNA and sat-RNA C and also to intensify symptoms of TCV infection. These recently characterized defective interfering (DI) RNAs are derived from the helper viral (TCV) RNA, contain the 5′ and 3′ terminal segments of the viral RNA, and interfere with helper viral replication in the plant. As noted above for the cucumoviral satellite RNAs, however, the search for pathogenicity determinants in the TCV system is likely to be more complex than initially imagined. First, mutations in the 5′ end of sat-RNA C have been shown to affect symptom expression (Simon et al. 1988), and, second, there is now clear evidence (Collmer and Howell 1990) that the presence of sat-RNA C in association with a distinct but very similar isolate of TCV (TCV-B) does not result in the same severe crinkling and stunting as originally described with the John Innes isolate.

Viral Proteins Associated with Symptom Induction

Viral-encoded proteins have also been implicated in symptom production in infected plants. The availability of cloned plant viral genomes from which infectious transcripts can be synthesized, coupled with the existing abundance of structural data on viral capsid proteins, promises rapid advances in this area of structure-function analysis.

The TMV Coat Protein is a Determinant of Host Response

Although observations over 100 years of working with TMV strains and mutants had implicated the viral coat protein in both the hypersensitive response in resistant tobacco species and differences in symptomatology in susceptible plants, work of only the last five years has demonstrated that the coat protein gene is a determinant of host response. Making use of infectious TMV cDNA clones, Saito et al. (1987) showed that exchange of the coat protein genes between two strains of TMV, which induced either localized necrotic lesions (hypersensitivity) or a systemic mosaic disease in plants with the *N′* gene, also reversed their phenotype. Knorr and Dawson (1988) went one step further to show that an experimentally induced point mutation within the U1-TMV coat protein gene, encoding a single amino acid change from serine to phenylalanine at position 148 of the coat protein, resulted in a change in host response in *N′* plants from susceptibility and mosaic symptom development to hypersensitivity and necrosis. Later work by Culver and Dawson (1989) showed that individual amino acid substitutions at four other places in the U1-TMV coat protein also led to the induction of hypersensitivity in these plants, thereby demonstrating that a number of sites in the coat protein gene of TMV affect the host response. It is an interesting aside to note that TMV mutants lacking a capsid gene still induce the hypersensitive response in tobacco carrying the alternative resistance gene *N*, a finding that indicates that different viral factors are involved in the induction of hypersensitivity with the two resistance genes.

Although the studies of TMV in tobacco containing the *N′* resistance gene implicate the coat protein gene in the induction of hypersensitivity, other work by Dawson et al. (1988) has shown that TMV mutants resulting from a series of deletions in the coat protein gene induce a range of differing symptoms in the susceptible tobacco variety Xanthi. Because the mutants all encode some form of truncated coat protein lacking the site essential for binding to viral

RNA, the resulting coat proteins produced in infected plants are not sequestered into virions. Thus, although all mutants seemed to replicate with similar efficiency, accumulation of coat protein and long distance spread was variable. Two interesting findings emerged from observations on the variety of different symptoms produced by the different mutants. First, and quite surprisingly, two of the mutants induced distinct necrotic lesions on inoculated leaves of Xanthi tobacco, a susceptible host that normally does not produce a hypersensitive response to TMV. This unexpected result demonstrated a previously unknown capacity in these plants to exhibit a hypersensitive resistance response to aberrant forms of the virus. Second, there was a correlation between the structure of the coat protein and the symptoms produced. Mutants without the normal C terminus of the protein were usually symptomless, whereas all those that produced the normal C terminal region induced some degree of yellowing. In addition, there was a correlation between the amount of mutant coat protein accumulated in cells and the amount of yellowing of leaves.

The TMV Coat Protein Accumulates in Chloroplasts

Because disease symptoms involving yellowing suggest some type of chloroplast dysfunction, it may be important to note the demonstration by Reinero and Beachy (1986) that coat protein of TMV accumulates in chloroplasts of systemically infected leaves. Additionally, chloroplasts in leaves infected with three severe strains of TMV contained at least ten times more coat protein than chloroplasts from leaves infected with TMV strain PV42, which induces no symptoms. Perhaps most interestingly, the chloroplast pool of coat protein not sequestered in the virus-like particles in the stroma was associated with the thylakoid membranes and determined to be free coat protein molecules. Since photosystem II was inhibited in plants infected with the severe but not the mild TMV strain, the authors hypothesized that such free coat protein might interact with membrane-bound thylakoid proteins and interfere with electron transport (Reinero and Beachy 1989). Given that the work of Dawson et al. (1988) also correlates free coat protein with leaf yellowing, it seems worthwhile to consider the suggestion by Atreya and Siegel (1989) that the presence of initiation sites for TMV coat protein encapsidation on certain chloroplast RNAs may serve to moderate pathogenesis by sequestering coat protein into noninfectious rods (pseudovirions) within the chloroplasts. Of the strains examined, a strain of TMV (U2) that induces mild symptoms had the greatest capacity to form pseudovirions (Rochon and Siegel 1984). This property of U2 may appear to break the previously mentioned correlation (Reinero and Beachy 1986) between symptom severity and the amount of TMV coat protein within chloroplasts. The important variable, however, may be the amount of *free* coat protein, not assembled in pseudovirions but associated with the thylakoid membranes. In such a hypothesis, assembled coat protein would be innocuous; free coat protein would be harmful. This type of analysis of U2-TMV coat protein in chloroplasts has not been reported.

Although studies on the interaction between TMV coat protein and chloroplasts may provide insight into how viruses cause disease, a more basic unresolved question is how the coat protein initially appears inside the chloroplast. Both genomic length TMV RNA (Schoelz and Zaitlin 1989) and TMV coat protein produced in transgenic plants (Osbourn et al. 1989) can enter chloroplasts *in vivo*. TMV coat protein is probably transported into the chloroplast in some manner different from that of the small subunit of ribulose bisphosphate carboxylase. (TMV coat protein lacks a transit peptide sequence.) The less likely possibility that it is translated from the genomic TMV RNA on chloroplast ribosomes has not been definitively eliminated.

An alternative hypothesis relating the presence of TMV coat protein in chloroplasts to the production of plant viral disease has been presented. Recent studies with TMV suggest that viral symptoms could result from the sequestering of specific plant messenger RNAs into virion-like structures, thus preventing their expression at a critical stage in plant development. Rochon and Siegel (1984) and Atreya and Siegel (1989) have shown that certain transcripts from chloroplast DNA, including the mRNA for the large subunit of ribulose bisphosphate carboxylase/oxygenase, contain one or more TMV-encapsidation initiation sites that could result in their assembly into pseudovirions within chloroplasts of TMV-infected plants. In a study testing whether encapsidation of plant mRNA could alter its expression, Sleat et al. (1988) engineered transgenic tobacco plants to produce mRNAs for chloramphenicol acetyltransferase (CAT) that contained a TMV-encapsidation initiation site. Infection of such plants with TMV led to the encapsidation of these mRNAs by TMV coat protein and a concomitant reduction in CAT activity, particularly in systemically infected, developing leaves. Thus, sequestration of plant mRNAs by encapsidation with TMV coat protein, at least in this case, did alter plant gene expression. No data currently exist that correlate pseudovirions with symptom induction, however, nor has it been demonstrated that a substantial portion of the chloroplast RNA pool is sequestered into pseudovirions. The alternative possibility that such viral-host interactions might instead moderate pathogenicity (Atreya and Siegel 1989) has been discussed above.

Gene VI of Cauliflower Mosaic Virus is a Symptom-producing Determinant

When viral genes have been implicated in disease production, it has been difficult to distinguish a direct effect of the gene in symptom induction from an indirect effect on viral replication or movement. The ability to express isolated viral genes in transgenic plants now makes that distinction possible, as has been accomplished recently with gene VI of the DNA virus cauliflower mosaic virus (CaMV). Evidence implicating gene VI in pathogenesis first came from mutagenesis studies by Daubert et al. (1983), in which small in-phase insertions (four amino acids) at two places within gene VI dramatically reduced the severity of disease symptoms. Both mutations also reduced viral replication, however. To locate other symptom-producing determinants, hybrid genomes were constructed between CaMV isolates with widely differing symptom-producing phenotypes. Determinants affecting different symptoms, such as stunting, timing, and spread, mapped widely in the CaMV genome (Stratford and Covey 1989). Leaf chlorosis and mottling and the ability to infect systemically solanaceous hosts mapped to gene VI, however (Daubert et al. 1984; Schoelz et al. 1986; Stratford and Covey 1989).

Several laboratories have constructed transgenic tobacco plants in which gene VI is expressed from a gene copy integrated into the chromosome (Goldberg et al. 1987; Baughman et al. 1988; Takahashi et al. 1989). In these studies, the appearance of viral-like symptoms was correlated with the levels of gene VI product, which varied from plant to plant. Symptoms resulted from the gene VI protein product itself and not from the RNA, since symptom production was blocked by either a deletion or a frame-shifting linker mutation in the gene. The plants used in these experiments were tobacco plants that are easily transformable but are not hosts for CaMV. Nonetheless, these nonhost transgenic plants displayed a viral infection syndrome, including chlorosis, growth suppression, and accumulation of pathogenesis-related proteins, characteristic of viral infection in host plants. As might be expected from the chlorotic symptoms observed on a macroscopic scale, electron micrographs show that the chloroplasts in the gene VI-expressing transgenic plants are disrupted (E. M. Herman and S. H. Howell, unpublished observations). An effort is under way to introduce gene VI, as a transgene, into normal CaMV hosts. The fact that transgenic plants of several species (tobacco, tomato, potato) that express the coat protein of different plant viruses (TMV, CMV, alfalfa mosaic virus, potato virus X) all have a normal phenotype (Powell Abel et al. 1986; Tumer et al. 1987; Cuozzo et al. 1988; Hoekema et al. 1989) suggests that symptom production in the gene VI-expressing plants is related specifically to the action of the viral protein itself rather than to the nonspecific expression of a virus-derived protein within plant cells.

The mechanism by which the gene VI produces disease symptoms is unknown. In the context of the virus, gene VI encodes a protein, p62, which is the major component of viral inclusion bodies found in the cytoplasm of infected cells, believed to be the site of viral reverse transcription. Recently, it has been demonstrated that the gene VI product also acts as a transactivator of CaMV and figwort mosaic virus (FMV) gene expression (Bonneville et al. 1989, Gowda et al. 1989). The gene VI product appears to operate at a post-transcriptional level promoting the expression of the tandemly arranged genes on the 35S RNA. Whether the ability of p62 to trans-activate viral gene expression is related to its ability to produce symptoms and alter gene expression in the host is unknown.

Conclusion and Summary

How viruses induce disease symptoms in plants has been a long-standing question. Only recently have new techniques in molecular biology presented the opportunity to analyze the role(s) in pathogenesis of specific domains of viral nucleic acids and individual viral proteins. For both small pathogenic RNAs (viroids and satellite RNAs) and plant viruses, the availability of infectious cDNA clones that can be altered *in vitro* has offered new possibilities to discover structure-function relationships in pathogenic molecules. In addition, the use of transgenic plants provides the opportunity to study the effects of the expression of a single viral gene product independent of its role in the infection. As a result, rapid progress should be made in delineating domains of model pathogenic molecules that serve as determinants of viral pathogenesis. On the other hand, how such molecules interact with plant cell metabolism remains obscure, one example being the unsolved puzzle of why mosaic- or mottle-like symptoms appear in tobacco plants transgenic for gene VI, where presumably every cell is expressing the gene product, p62. With the power of the new technologies now focused resolutely on the plant cell (for example, Young et al. 1992, this volume), however, target molecules with which viral molecules interact inevitably will be found. As the role of these target molecules in normal plant metabolism is unraveled, an increased understanding of mechanisms of plant viral pathogenesis will follow.

Literature Cited

Altenbach, S. B., and S. H. Howell. 1981. Identification of a satellite RNA associated with turnip crinkle virus. Virology 112:25–33.

Atreya, C. D., and A. Siegel. 1989. Localization of multiple TMV encapsidation initiation sites on *rbcL* gene transcripts. Virology 168:388–392.

Baughman, G. A., J. D. Jacobs, and S. H. Howell. 1988. Cauliflower mosaic virus gene VI produces a symptomatic phenotype in transgenic tobacco plants. Proc. Natl. Acad. Sci. USA 85:733–737.

Bonneville, J. M., H. Sanfacon, J. Fütterer, and T. Hohn. 1989. Posttranscriptional *trans*-activation in cauliflower mosaic virus. Cell 59:1135–1143.

Carrington, J. C., T. J. Morris, P. G. Stockley, and S. C. Harrison. 1987. Structure and assembly of turnip crinkle virus. IV. Analysis of the coat protein gene and implications of the subunit primary structure. J. Mol. Biol. 94:265–276.

Collmer, C. W., and S. H. Howell. 1990. Both helper virus and satellite RNA affect symptoms of turnip crinkle virus infection. Phytopathology 80:983 (Abstr.)

Culver, J. N., and W. O. Dawson. 1989. Point mutations in the coat protein gene of tobacco mosaic virus induce hypersensitivity in *Nicotiana sylvestris*. Molec. Plant-Microbe Interact. 2:209–213.

Cuozzo, M., K. M. O'Connell, W. Kaniewski, R.-X. Fang, N.-H. Chua, and N. E. Tumer. 1988. Viral protection in transgenic tobacco plants expressing the cucumber mosaic virus coat protein or its antisense RNA. Bio/Technology 6:549–557.

Daubert, S. 1988. Sequence determinants of symptoms in the genomes of plant viruses, viroids, and satellites. Molec. Plant-Microbe Interact. 1:317–325.

Daubert, S. D., J. Schoelz, L. Debao, and R. J. Shepherd. 1984. Expression of disease symptoms in cauliflower mosaic virus genomic hybrids. J. Mol. Appl. Genet. 2:537–547.

Daubert, S., R. J. Shepherd, and R. C. Gardner. 1983. Insertional mutagenesis of the cauliflower mosaic virus genome. Gene 25:201–208.

Dawson, W. O., P. Bubrick, and G. L. Grantham. 1988. Modifications of the tobacco mosaic virus coat protein gene affecting replication, movement, and symptomatology. Phytopathology 78:783–789.

Devic, M., M. Jaegle, and D. Baulcombe. 1989. Symptom production on tobacco and tomato is determined by two distinct domains of the satellite RNA of cucumber mosaic virus (strain Y). J. Gen. Virol. 70:2765–2774.

Fraser, R. S. S. 1985. Mechanisms of Resistance to Plant Diseases. Dordrecht, The Netherlands: Martinus Nijhoff/Dr. W. Junk.

Fraser, R. S. S. 1987. Biochemistry of Virus-Infected Plants. Letchworth, England: Research Studies Press Ltd.

Fraser, R. S. S., and S. A. R. Loughlin. 1980. Resistance to tobacco mosaic virus in tomato: Effects of the *Tm-1* gene on virus multiplication. J. Gen. Virol. 48:87–96.

Goldberg, K. B., M. J. Young, J. E. Schoelz, J. M. Kiernan, and R. J. Shepherd. 1987. Single gene of CaMV induces disease. Phytopathology 77:1704 (Abstr.).

Gowda, S., F. C. Wu, H. B. Scholthof, and R. J. Shepherd. 1989. Gene VI of figwort mosaic virus (caulimovirus group) functions in posttranscriptional expression of genes on the full-length RNA transcript. Proc. Natl. Acad. Sci. USA 86:9203–9207.

Haas, B., A. Klanner, K. Ramm, and H. L. Sänger. 1988. The 7S RNA from tomato leaf tissue resembles a signal recognition particle RNA and exhibits a remarkable sequence complementarity to viroids. EMBO J. 7:4063–4074.

Harders, J., N. Lukacs, M. Robert-Nicoud, T. M. Jovin, and D. Riesner. 1989. Imaging of viroids in nuclei from tomato leaf tissue by *in situ* hybridization and confocal laser scanning microscopy. EMBO J. 8:3941–3949.

Hoekema, A., M. J. Huisman, L. Molendijk, P. J. M. van den Elzen, and B. J. C. Cornelissen. 1989. The genetic engineering of two commercial potato cultivars for resistance to potato virus X. Bio/Technology 7:273–278.

Jacquemond, M., and G. J.-M. Lauquin. 1988. The cDNA of cucumber mosaic virus-associated satellite RNA has *in vivo* biological properties. Biochem. Biophys. Res. Commun. 151:388–395.

Kaper, J. M., and C. W. Collmer. 1988. Modulation of viral plant diseases by secondary RNA agents. In RNA Genetics, Vol. III, pp. 171–194. Eds. E. Domingo, J. Holland and P. Ahlquist. Boca Raton, FL: CRC Press.

Kaper, J. M., M. E. Tousignant, and L. M. Geletka. 1990. Cucumber mosaic virus-associated RNA 5. XII. Symptom modulating effect is codetermined by the helper virus satellite replication support function. Res. Virol. 141:487–503.

Kaper, J. M., M. E. Tousignant, and M. T. Steen. 1988. Cucumber mosaic virus-associated RNA 5. XI. Comparison of 14 CARNA 5 variants relates ability to induce tomato necrosis to a conserved nucleotide sequence. Virology 163:284–292.

Kaper, J. M., and H. E. Waterworth. 1977. Cucumber mosaic virus associated RNA 5: Causal agent for tomato necrosis. Science 196:429–431.

Knorr, D. A., and W. O. Dawson. 1988. A point mutation in the tobacco mosaic virus capsid protein gene induces hypersensitivity in *Nicotiana sylvestris*. Proc. Natl. Acad. Sci. USA 85:170–174.

Kurath, G., and P. Palukaitis. 1989. Satellite RNAs of cucumber mosaic virus: Recombinants constructed *in vitro* reveal independent functional domains for chlorosis and necrosis in tomato. Molec. Plant-Microbe Interact. 2:91–96.

Li, X. H., L. A. Heaton, T. J. Morris, and A. E. Simon. 1989. Turnip crinkle virus defective interfering RNAs intensify viral symptoms and are generated *de novo*. Proc. Natl. Acad. Sci. USA 86:9173–9177.

Li, X. H., and A. E. Simon. 1990. Symptom intensification on cruciferous hosts by the virulent satellite RNA of turnip crinkle virus. Phytopathology

80:238–242.
Masuta, C., S. Kuwata, and Y. Takanami. 1988. Disease modulation on several plants by cucumber mosaic virus satellite RNA (Y strain). Ann. Phytopathol. Soc. Jpn. 54:332–336.
Masuta, C., and Y. Takanami. 1989. Determination of sequence and structural requirements of a cucumber mosaic virus satellite RNA (Y-satRNA). Plant Cell 1:1165–1173.
Matthews, R. E. F. 1973. Induction of disease by viruses, with special reference to turnip mosaic virus. Ann. Rev. Phytopathol. 11:147–170.
Osbourn, J. K., K. A. Plaskitt, J. W. Watts, and T. M. A. Wilson. 1989. Tobacco mosaic virus coat protein and reporter gene transcripts containing the TMV origin-of-assembly sequence do not interact in double-transgenic tobacco plants: Implications for coat protein-mediated protection. Molec. Plant-Microbe Interact. 2:340–345.
Palukaitis, P. 1988. Pathogenicity regulation by satellite RNAs of cucumber mosaic virus: Minor nucleotide sequence changes alter host responses. Molec. Plant-Microbe Interact. 1:175–181.
Powell Abel, P., R. S. Nelson, B. De, N. Hoffman, S. G. Rogers, R. T. Fraley, and R. N. Beachy. 1986. Delay of disease development in transgenic plants that express the tobacco mosaic virus coat protein gene. Science 232:738–743.
Reinero, A., and R. N. Beachy. 1986. Association of TMV coat protein with chloroplast membranes in virus-infected leaves. Plant Mol. Biol. 6:291–301.
Reinero, A., and R. Beachy. 1989. Reduced photosystem II activity and accumulation of viral coat protein in chloroplasts of leaves infected with tobacco mosaic virus. Plant Physiol. 89:111–116.
Rochon, D., and A. Siegel. 1984. Chloroplast DNA transcripts are encapsidated by tobacco mosaic virus coat protein. Proc. Natl. Acad. Sci. USA 81:1719–1723.
Saito, S., T. Meshi, N. Takamatsu, and Y. Okada. 1987. Coat protein gene sequence of tobacco mosaic virus encodes a host response determinant. Proc. Natl. Acad. Sci. USA 84:6074–6077.
Schnölzer, M., B. Haas, K. Ramm, H. Hofmann, and H. L. Sänger. 1985. Correlation between structure and pathogenicity of potato spindle tuber viroid (PSTV). EMBO J. 4:2181–2190.
Schoelz, J., R. J. Shepherd, and S. Daubert. 1986. Region VI of cauliflower mosaic virus encodes a host range determinant. Mol. Cell. Biol. 6:2632–2637.
Schoelz, J. E., and M. Zaitlin. 1989. Tobacco mosaic virus RNA enters

chloroplasts *in vivo*. Proc. Natl. Acad. Sci. USA 86:4496–4500.

Simon, A. 1988. Satellite RNAs of plant viruses. Plant Mol. Biol. Rptr. 6:240–252.

Simon, A., H. Engel, R. P. Johnson, and S. H. Howell. 1988. Identification of regions affecting virulence, RNA processing and infectivity in the virulent satellite of turnip crinkle virus. EMBO J. 7:2645–2651.

Simon, A., and S. H. Howell. 1986. The virulent satellite RNA of turnip crinkle virus has a major domain homologous to the 3′ end of the helper virus genome. EMBO J. 5:3423–3428.

Sleat, D. E., K. A. Plaskitt, and T. M. A. Wilson. 1988. Selective encapsidation of CAT gene transcripts in TMV-infected transgenic tobacco inhibits CAT synthesis. Virology 165:609–612.

Stratford, R., and S. N. Covey. 1989. Segregation of cauliflower mosaic virus symptom genetic determinants. Virology 172:451–459.

Takahashi, H., K. Shimamoto, and Y. Ehara. 1989. Cauliflower mosaic virus gene VI causes growth suppression, development of necrotic spots and expression of defence-related genes in transgenic tobacco plants. Mol. Gen. Genet. 216:188–194.

Tumer, N. E., K. M. O'Connell, R. S. Nelson, P. R. Sanders, R. N. Beachy, R. T. Fraley, and D. M. Shah. 1987. Expression of alfalfa mosaic virus coat protein gene confers cross-protection in transgenic tobacco and tomato plants. EMBO J. 6:1181–1188.

Young, N. D., Messeguer, R., Golemboski, D. B., Tanksley, S. D. 1993. DNA markers for viral resistance genes in tomato. Resistance to Viral Diseases of Vegetables: Genetics and Breeding. Ed. M. M. Kyle. Portland, OR: Timber Press.

CHAPTER 12

Conclusions: Future Prospects, Strategies, and Problems

JOHN C. SORENSON, ROSARIO PROVVIDENTI & HENRY M. MUNGER

The preceding chapters present a general consensus that heritable resistance is the most effective and economical way to control plant viral diseases. Any successful viral resistance program should be based upon an understanding of the nature of this resistance and its limitations. Resistance can take several forms, including hypersensitivity, tolerance, resistance to viral multiplication and/or movement, and extreme resistance or immunity. Many of the available sources of resistance have proven to be monogenically inherited, thus facilitating the incorporation of resistance genes into commercial varieties. In addition, experience has shown that most of the genes for resistance to viruses are rather stable. A number of resistance genes have been successfully used for decades with great economic benefit to growers and consumers.

Similar to interactions between plants and fungal and bacterial pathogens, resistance to viruses can be isolate-specific. This

Dr. Sorenson is with the Asgrow Seed Co., Kalamazoo, MI 49001. Dr. Provvidenti is in the Department of Plant Pathology, Cornell University, New York State Agricultural Experiment Station, Geneva, New York 14456. Dr. Munger is in the Departments of Plant Breeding and Vegetable Crops, New York State College of Agriculture and Life Sciences, Cornell University, Ithaca, NY 14853.

specificity of resistance is an important factor to be considered in developing virus-resistant varieties because the performance of the varieties will ultimately depend upon the presence and distribution of particular isolates in a given area. In such cases it is necessary to pyramid all of the available resistance genes to make varieties more versatile. The number of viral pathotypes that must be considered is often not as large as for physiological races of many fungi and pathovars of many bacteria.

The authors of the preceding chapters also agree that most of the sources of resistance that we utilize in breeding programs come from foreign germplasm and that we are increasingly relying upon wild relatives of cultivated species for viral resistance genes. We are fortunate to have in the USDA germplasm resources a collection of domestic and foreign plant species that offers great genetic diversity. Nonetheless, though some crop species are very rich in wild relatives, others are seriously under-represented. It is therefore essential to continue to collect and preserve all the germplasm that is presently available because of the continuous destruction of plant habitats.

In the United States and abroad, in both private and public institutions, there are many germplasm collections. It is imperative that the integrity of these collections be maintained to serve as sources of genetic variability for use when new viruses or pathotypes are identified. Resistance genes should be identified before a viral disease reaches catastrophic proportions, given that the incorporation of resistance into horticulturally acceptable varieties requires several years of intensive breeding.

Because of the global nature of current agricultural and produce supply systems, plant viruses can travel great distances in a relatively short time. Improved communications about the appearance of new viral diseases and sources of resistance have enhanced the effectiveness of coordinated efforts among plant pathologists and breeders of various nations in solving common problems.

Great progress has been made in breeding better and more productive crops. Even a cursory look at the science of plant breeding reveals a tremendous evolution and refinement of the techniques available under the umbrella of variety development. Although biotechnology is often set apart from "conventional" plant breeding, it is becoming a logical and useful additional tool at the disposal of plant breeders. Biotechnology alone is unlikely to create useful varieties unless it is combined with a number of procedures used in conventional breeding; for example, setting realistic and needed objectives for improvement of a crop based on knowledge of the crop and its germplasm resources, selecting among and within the variant forms produced by advanced technologies, evaluating prospective varieties, multiplying seed, etc. Although the highly glamorized approaches to crop improvement have not yet been commercialized, it is becoming clear that they will make substantial contributions to variety development. Biotechnology techniques have already achieved dramatic progress in the area of viral resistance. A comprehensive review of the progress and techniques used to engineer viral resistance is beyond the scope of this brief chapter; however, we briefly describe some of the highlights in this promising area. Grumet (1990) has recently published a more detailed review of the subject.

A number of approaches to viral resistance involve biotechnology. Embryo culture, which might be considered a forerunner of biotechnology, has been used with considerable success since about 1940 to overcome fertility barriers in wide crosses (Smith 1944), and it has made possible the transfers of several viral resistances from related species. With improved methodology, the technique has been renamed "embryo rescue" and will contribute even more to viral resistance breeding.

An additional aid to the plant breeder, which holds considerable potential for speeding the introgression process, involves DNA fingerprinting techniques commonly referred to as restriction fragment length polymorphisms (RFLPs) (see Young et al. 1993, this volume). Once a correlation between a particular DNA fingerprint and a resistance trait has been established, the breeder can use this laboratory technique to follow the presence of the resistance gene among backcross progeny. This can be a particularly powerful approach when the off-season greenhouse screening for the resistance gene by traditional inoculation tests is either difficult or unreliable. Although this approach is still not developed sufficiently for utilization by plant breeders in most species, it is clear that the potential for increased breeding efficiency is very real.

A number of molecular strategies to developing new sources of viral genes are beginning to emerge. Only a few of these have been developed to the point that their potential utility can be adequately evaluated, but the array of new tactics is quite impressive. Equally interesting is that most of these approaches do not involve the identification and isolation of an existing resistance gene. One particularly attractive aspect of many of the new molecular approaches is that genes can, theoretically, be transferred into any plant species without the limitation of fertility barriers. Genes of various types have been successfully transferred into plant chromosomes from bacteria, fungi, animals, viruses, and an extremely diverse array of other plant species. In each of these cases, potentially useful traits have been effectively expressed in the recipient plants. The degree of expression of a gene may differ considerably when transferred to a new genetic background, however. Even within the narrower limits of transfer by conventional methods, some genes for viral resistance have been less effective when moved into a different species.

The genetic engineering approach to viral resistance that has been the best studied to date involves a phenomenon known as coat protein–derived cross protection, coat protein–mediated resistance, or some variation on the theme. Although coat protein–derived resistance and classical cross protection share a number of similarities, it has not been well documented whether the two operate through similar mechanisms. In the coat protein–derived resistance approach, the gene encoding a primary coat protein or a capsid component is isolated from the virus in question and inserted into the chromosomes of the susceptible plant. For a number of plant viral groups in a variety of host species, expression of the transgene confers some level of protection against infection by the virus used as the donor of the gene. The approach is particularly attractive because the viral genome is quite simple and the coat protein genes are relatively easy to identify and isolate. Experiments have shown significant variability in the level of protection observed in these plants, ranging from a brief delay in the appearance of symptoms (relative to nontransgenic control plants) to very high levels of resistance approaching immunity. There does appear to be stronger protection within some virus groups than others, but the information base is still too limited to generalize about effectiveness. In addition, even within a series of transgenic plants carrying the coat protein gene for a given virus, one sees a substantial range of protection levels. This range is commonly observed when foreign genes are inserted into the chromosomes of a plant and is most probably caused by insertion of the genes into chromosomal locations that are more or less conducive to expression of the gene. Thus, a part of any successful transformation strategy must be to generate a large number of transformants and select those which exhibit a high level of expression and/or protection. Inferring efficacy from one or a few independent transformation events is inappropriate.

Cumulative data on the effectiveness of coat protein–mediated resistance across several virus groups in several plant species suggest that this approach will not provide the final answer for viral protection. Although there are reports of near immunity resulting from the technique, in general the level of protection is less than would be desirable. The protection is typically overcome by high levels of inoculum, and the protection is often manifest as a delay in the development of symptoms rather than a prevention of symptom development. Variety developers are accustomed to imperfect or incomplete protection, however, and abandoning the approach this early in its development would be inappropriate. Several seed companies are attempting to develop and commercialize cucurbit varieties that will carry the coat protein genes for four viruses that cause severe economic loss to growers in the primary cucurbit production areas. (Some of the factors involved in making the decision to proceed along these lines are discussed below.) Although protection is incomplete based on a limited series of recent field trials, the resistance is significant and likely to provide substantial benefit to growers who may otherwise be driven out of production. Conventional sources of resistance are being incorporated into commercially acceptable varieties in several publicly funded breeding programs. The real solution to the virus problems in cucurbits will probably involve "piggybacking" two or more types of resistance. Plant breeders have long understood the benefits of incorporating major gene resistance into backgrounds carrying a significant level of field tolerance. Coat protein–mediated resistance may well provide a useful companion for conventional sources of resistance.

A second biotechnology-based approach to virus control involves "antisense" technology. This approach, which has numerous potential applications beyond virus control, is based on the strand complementarity of nucleic acids. The majority of plant viruses have single-stranded (+) sense RNA genomes. The concept behind antisense is to insert a gene into the host plant that encodes an RNA complementary to a critical region of the viral genome. This "antisense" RNA will pair with the viral RNA upon infection and uncoating to physically block replication and/or translation of the viral genome. Attempts to utilize this approach have been disappointing. Although some protection has been observed, it has been quite weak and appears to have little commercial potential in its present state of development. Another phenomenon, termed co-suppression, involves the apparent repression of endogenous genes as a consequence of transformation with similar sequences. The basis for this effect of suppression of genes by similar "sense" sequences is unknown but demonstrates the complexity of interactions affecting gene expression in transgenic systems. This technology is in the earliest stages of its development, however, and improvements in the process may eventually make it a viable strategy in the future.

A third approach to virus control is the use of attenuating satellites or defective interfering RNAs (DIs). Some classes of viruses may contain satellites or DIs that attenuate (but in some cases enhance) the severity of the viral symptoms. The concept behind the approach is to transfer the satellite sequences into plant chromosomes. During an infection with the virus, the satellite sequences are amplified by the viral replicative machinery. It is presumed that the satellite sequences compete with the legitimate viral RNA for replicative complexes and for coat proteins, thereby reducing the rate and extent of virus replication and spread. There is a significant concern with this approach, however, resulting from our poor understanding of satellite behavior. Some satellites actually enhance virulence, and in some cases the same satellite may attenuate one strain of a virus and enhance another. Obviously, the risks inherent in such an approach preclude its usefulness until we have a better understanding of the dynamics of satellites and their role in viral etiology.

One relative newcomer on the scene has generated considerable interest. This approach involves the transfer of an altered viral replicase gene into the host plants (Golemboski et al. 1990). When a gene from tobacco mosaic virus (which probably codes for a component of the viral replication complex but may simply compete with legitimate replicase components) is expressed in tobacco plants, these plants become highly resistant to infection by the virus. The level of this engineered resistance seems to be much higher than coat protein–mediated resistance. Work on this general phenomenon is ongoing in several laboratories with other viruses. If widely applicable, it could prove to be an important tool for engineering viral resistance.

These exciting advances are likely to bring together biotechnologists and plant breeders in a genuine partnership. A number of factors may disrupt the entrance of the new technologies into the marketplace, however. Presently, the regulatory path is extremely uncertain. It appears that plants carrying genetically engineered viral resistance would be subject to regulation by the USDA, the EPA, and the FDA. USDA guidelines are currently the best developed, but still do not give clear direction on the criteria for unrestricted (commercial) release. The EPA is still framing its guidelines, and apparently the FDA is even less prepared to provide the regulatory direction needed for commercialization of a genetically engineered plant. The issue is complicated further with respect to foreign activities. Much of the commercial production of vegetable seed is done in off-shore locations such as Thailand, Taiwan, India, and several South American countries. The governments of many of these countries have not yet begun to develop policies and guidelines for the importation and propagation of transgenic plants.

Beyond the issue of regulation, there are concerns about the public acceptance of genetically engineered food products. In the tens of thousands of person years of exposure to recombinant DNA, not a single documented case of exposure-related illness has been reported. It is clear beyond any rational protest that there is no inherent public health risk associated with recombinant DNA *per se*. Unfortunately, public perceptions are not always rational, as some of the recent food safety initiatives have demonstrated. The value of genetic resistance (including genetically engineered resistance) in reducing chemical usage seems obvious to those engaged in the area of research. Public opinion polls have shown repeatedly, however, that the advantages of genetic engineering are not obvious to most members of the nonscientific community, and that these solutions are not universally viewed as acceptable alternatives. These are sobering factors that must be considered as elements of risk by any company intending to commercialize the products of this research.

Decisions about investing in biotechnology have proven to be vexing ones for most seed companies. The seed industry has enjoyed a long period of reasonably low risk in the research investments it has made. The science has had a significant "reliability" based on many years of practical application. Breeding is of course not always predictable, but commercial seed companies have generally felt comfortable in weighing the effectiveness of breeding technology in developing new varieties against the costs of doing the research. Beyond the comfort resulting from the familiarity of the technology, the seed industry has also benefitted from a strong and effective public breeding effort that has assumed the riskiest stages of the process. Most of the innovations in plant breeding have evolved from within the discipline or in allied fields. Even the newest tools in the arsenal of the plant breeder seem to be logical extensions of the science rather than radical new approaches.

Biotechnology has introduced a new element into the decisions a company must make about research investments. For the first time, seed companies have been forced to make investment decisions with respect to technology that has not yet been proven to have any value in new product development. Several technological developments seem to assure that biotechnology will have value in developing new varieties, but none have yet been through the trial of commercialization. The more difficult issues for seed companies, however, are not if biotechnology will contribute to variety development, but when and how much investment will be necessary to exploit the tool effectively. Given that the resources that can be applied to research are limited and that investments in biotechnology will be made at the expense of investments in more established technologies, research directors are faced with risk-balancing decisions having a far greater level of uncertainty than ever before. Since the cost of investing in biotechnology is very large, these decisions will also have a much larger impact on the future health of individual companies than in the past.

One need only examine the levels of biotechnology investment currently being made by various companies to see that the industry has no one best guess as to what the optimal balance may be. Most vegetable seed companies have elected to take a somewhat conservative approach to this problem.

Biotechnology will probably become a very significant factor in the development of varieties in the future. No major seed company is likely to thrive without effectively integrating biotechnology into its product development process. Biotechnology will be only one of the many tools that a variety development program will need to use effectively, however, and it is not even likely to be the most important one. Plant breeding will continue to be the central engine that drives variety development. Biotechnology will probably accomplish specific discrete jobs for the plant breeder, but currently it lacks the breadth, range, and sustained power to diminish the role of plant breeding in the highly complex process of variety development.

Use of conventional, cellular, and molecular approaches in parallel does not imply that they are competing technologies. In fact, the second phase of combining conventionally and biotechnologically derived resistance is well under way. Each approach has its strengths and liabilities. The key to future success in variety development will be to apply the strengths of each to compensate for inherent weaknesses. Through biotechnology, it may be possible to convert an entire product line to virus resistance in a relatively brief period of time, albeit with an imperfect resistance with an unknown degree of "durability." A conventional program has the potential to provide resistant materials in a limited number of types and a limited range of maturities in a similar time frame. Use of the two approaches separately is designed to provide an acceptable range of varieties to cover the needs of diverse geographic and market segments until they can be combined to provide a level of protection superior to either alone. An important requirement is that virus-resistant products, however they are developed, must be acceptable in other respects to growers, marketers, and consumers. The level of horticultural acceptability required may well vary with the severity of the virus problems. The resistant products that succeed will be those that provide the best value to the grower, shipper, retailer, and consumer. The primary determinants will again be the complex combinations of factors that only the well-skilled plant breeder can assemble.

We are privileged to be participants in the fusion of "conventional" breeding and the new technologies. The explosion of basic science that is accompanying this fusion will allow understanding of biological processes to an unprecedented extent. The greater impact of biotechnology may likely result from what we learn about processes rather than what we commercialize directly. In no area will that process be any more dynamic than in the interaction of hosts and their pathogens. Future developments on this topic should be exciting, indeed!

Literature Cited

Golemboski, D. B., Lomonossoff, G. P., and Zaitlin, M. 1990. Plants transformed with a tobacco mosaic virus nonstructural gene sequence are resistant to the virus. Proc. Natl. Acad. Sci. USA 87:6311–6315.

Grumet, R. 1990. Genetically engineered plant virus resistance. Hortscience 25(5):508–513.

Smith, P. G. 1944. Embryo culture of a tomato species hybrid. Proc. Amer. Hort. Sci. 44:413–416.

Young, N. D., R. Messeguer, D. Golemboski, and S. D. Tanksley. 1993. DNA markers for viral resistance genes in tomato. In Resistance to Viral Diseases of Vegetables: Genetics and Breeding. Ed. M. M. Kyle. Portland, OR: Timber Press.

Contributing Authors and Reviewers

Authors

C. W. Collmer
Department of Biology, Wells College, Aurora, NY 13026

W. T. Federer
Biometrics Unit, Department of Plant Breeding and Biometry, Cornell University, Ithaca, NY 14853

D. B. Golemboski
Department of Plant Pathology, Cornell University, Ithaca, NY 14853

S. M. Gray
United States Department of Agriculture/Agricultural Research Service and Department of Plant Pathology, Cornell University, Ithaca, NY 14853

S. H. Howell
Plant Molecular Biology Program, Boyce Thompson Institute, Ithaca, NY 14853

M. M. Kyle
Department of Plant Breeding and Biometry, Cornell University, Ithaca, NY 14853

R. Messeguer
Department of Plant Breeding and Biometry, Cornell University, Ithaca, NY 14853

J. W. Moyer
Department of Plant Pathology, North Carolina State University, Raleigh, NC 27695

H. M. Munger
Departments of Plant Breeding and Biometry and Vegetable Crops, Cornell University, Ithaca, NY 14853

R. Provvidenti
Department of Plant Pathology, New York State Agricultural Experiment Station, Cornell University, Geneva, NY 14456

R. W. Robinson
Department of Horticultural Sciences, New York State Agricultural Experiment Station, Cornell University, Geneva, NY 14456

B. T. Scully
Everglades Research and Education Center, IFAS, University of Florida, Belle Glade, FL 33430

J. C. Sorenson
Asgrow/Upjohn Inc., Kalamazoo, MI 49001

T. H. Superak
Harris Moran Seed Co., Sun City, FL 33586

S. D. Tanksley
Department of Plant Breeding and Biometry, Cornell University, Ithaca, NY 14853

J. C. Watterson
Petoseed Co., Inc., Woodland, CA 95695

N. D. Young
Department of Plant Pathology, University of Minnesota, St. Paul, MN 55108

Reviewers

L. L. Black
Department of Plant Pathology and Crop Physiology, Lousiana State University, Baton Rouge, LA 70803

W. R. Coffman
Department of Plant Breeding and Biometry, Cornell University, Ithaca, NY 14853

C. W. Collmer
Department of Biology, Wells College, Aurora, NY 13026

E. L. Cox
Asgrow Technical Center, El Centro, CA 92244

D. P. Coyne
Department of Horticulture, University of Nebraska, Lincoln, NE 68588

J. R. Dunlap
Texas Agricultural Experiment Station, Texas A&M University, Weslaco, TX 78399

D. W. Groff
Asgrow Seed Co., Tifton, GA 331794

A. O. Jackson
Department of Plant Pathology, University of California, Berkeley, CA 94720

J. D. McCreight
United States Department of Agriculture/Agricultural Research Service, Salinas, CA 93905

M. E. Miller
Texas Agricultural Experiment Station, Texas A&M University, Weslaco, TX 78399

M. A. Mutschler
Department of Plant Breeding and Biometry, Cornell University, Ithaca, NY 14853

R. Provvidenti
Department of Plant Pathology, New York State Agricultural Experiment Station, Cornell University, Geneva, NY 14456

E. J. Ryder
United States Department of Agriculture Agricultural Research Service, Salinas, CA 93905

S. J. Schwager
Biometrics Unit, Department of Plant Breeding and Biometry, Cornell University, Ithaca, NY 14853

B. T. Scully
Everglades Research and Education Center, IFAS, University of Florida, Belle Glade, FL 33420

M. E. Sorrells
Department of Plant Breeding and Biometry, Cornell University, Ithaca, NY 14853

E. J. Souza
University of Idaho, Aberdeen, ID 83210

B. Villalon
Texas Agricultural Experiment Station, Texas A&M University, Weslaco, TX 78399

D. H. Wallace
Departments of Plant Breeding and Biometry and Vegetable Crops, Cornell University, Ithaca, NY 14853

J. C. Watterson
Petoseed Co., Inc., Woodland, CA 95695

M. Zaitlin
Department of Plant Pathology, Cornell University, Ithaca, NY 14853

F. W. Zink
Department of Vegetable Crops, University of California, Davis, CA 95616

T. A. Zitter
Department of Plant Pathology, Cornell University, Ithaca, NY 14853

Index